Ben Stacy Jerrik (Ed.)

2695 Christabel

Ben Stacy Jerrik (Ed.)

2695 Christabel

Asteroid belt, Asteroid, Edward L. G. Bowell, Anderson Mesa Station

Part Press

Imprint

Publisher:
Part Press is a trademark of
International Book Market Service Ltd., 17 Rue Meldrum, Beau Bassin, 1713-01 Mauritius
Email: info@bookmarketservice.com
Website: www.bookmarketservice.com

Published in 2013

Printed in: U.S.A., U.K., Germany. This book was not produced in Mauritius.

ISBN: 978-613-9-36072-7

Contents

Articles

2695_Christabel 1
Asteroid_belt 2
Asteroid 16
Edward_L._G._Bowell 33
Anderson_Mesa_Station 33
Minor_planet 35
List_of_minor_planets 40
Small_Solar_System_body 43
Orbital_eccentricity 45
Orbital_period 48
Mean_anomaly 51
Orbital_inclination 52
Longitude_of_the_ascending_node 55

References

Article Sources and Contributors 57
Image Sources, Licenses and Contributors 59

2695_Christabel

Discovery and designation	
Discovered by	Bowell, E.
Discovery site	Flagstaff (AM)
Discovery date	October 17, 1979
Designations	
MPC designation	2695
Pronunciation	I
Alternative names	1979 UE
Orbital characteristics	
Epoch May 14, 2008	
Ap	2.9208551
Peri	2.4969549
Eccentricity	0.0782420
Orbital period	1628.5049894
Mean anomaly	185.64136
Inclination	14.87458
Longitude of ascending node	63.16177
Argument of peri	289.89321
Physical characteristics	
Geometric albedo	0.0995
Absolute magnitude *(H)*	12.30

2695 Christabel (1979 UE) is a main-belt asteroid discovered on October 17, 1979 by Bowell, E. at Flagstaff (AM).

External links

- JPL Small-Body Database Browser on 2695 Christabel (http://ssd.jpl.nasa.gov/sbdb.cgi?sstr=2695+Christabel)

Asteroid_belt

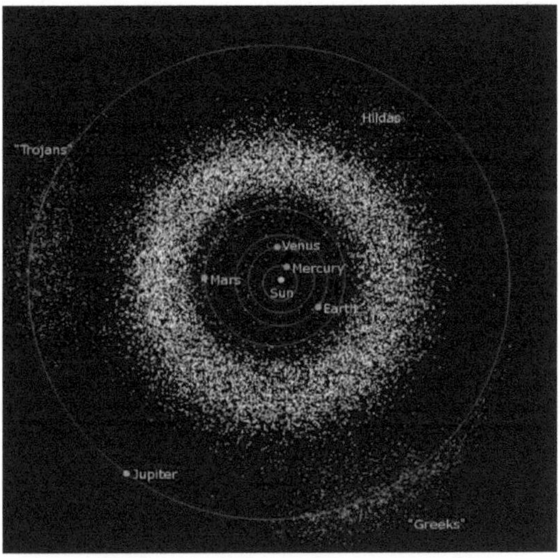

The asteroid belt (shown in white) is located between the orbits of Mars and Jupiter.

The **asteroid belt** is the region of the Solar System located roughly between the orbits of the planets Mars and Jupiter. It is occupied by numerous irregularly shaped bodies called asteroids or minor planets. The asteroid belt is also termed the **main asteroid belt** or **main belt** to distinguish its members from other asteroids in the Solar System such as near-Earth asteroids and trojan asteroids. About half the mass of the belt is contained in the four largest asteroids, Ceres, Vesta, Pallas, and Hygiea. These have mean diameters of more than 400 km, while Ceres, the asteroid belt's only dwarf planet, is about 950 km in diameter.[1] [2] [3] [4] The remaining bodies range down to the size of a dust particle. The asteroid material is so thinly distributed that numerous unmanned spacecraft have traversed it without incident. Nonetheless, collisions between large asteroids do occur, and these can form an asteroid family whose members have similar orbital characteristics and compositions. Nesvorny and Jenniskens (2010 Astrophysical Journal) attributed 85 percent of the Zodiacal Light dust to fragmentations of Jupiter-family comets, rather than from comets and collisions between asteroids in the asteroid belt. It was once thought that collisions of asteroids produce a fine dust that forms a major component of the zodiacal light. Individual asteroids within the asteroid belt are categorized by their spectra, with most falling into three basic groups: carbonaceous (C-type), silicate (S-type), and metal-rich (M-type).

The asteroid belt formed from the primordial solar nebula as a group of planetesimals, the smaller precursors of the planets, which in turn formed protoplanets. Between Mars and Jupiter, however, gravitational perturbations from the giant planet imbued the protoplanets with too much orbital energy for them to accrete into a planet. Collisions became too violent, and instead of fusing together, the planetesimals and most of the protoplanets shattered. As a result, most of the asteroid belt's mass has been lost since the formation of the Solar System. Some fragments can eventually find their way into the inner Solar System, leading to meteorite impacts with the inner planets. Asteroid orbits continue to be appreciably perturbed whenever their period of revolution about the Sun forms an orbital resonance with Jupiter. At these orbital distances, a Kirkwood gap occurs as they are swept into other orbits.

Classes of small Solar System bodies in other regions include the centaurs, Kuiper belt and scattered disk objects, and Oort cloud comets.

History of observation

Giuseppe Piazzi, discoverer of Ceres, known as a planet for many years, then as asteroid number 1, and eventually, a dwarf planet

In an anonymous footnote to his 1766 translation of Charles Bonnet's *Contemplation de la Nature*,[5] the astronomer Johann Daniel Titius of Wittenberg[6] [7] noted an apparent pattern in the layout of the planets. If one began a numerical sequence at 0, then included 3, 6, 12, 24, 48, etc., doubling each time, and added four to each number and divided by 10, this produced a remarkably close approximation to the radii of the orbits of the known planets as measured in astronomical units. This pattern, now known as the Titius–Bode law, predicted the semi-major axes of the six planets of the time (Mercury, Venus, Earth, Mars, Jupiter and Saturn) provided one allowed for a "gap" between the orbits of Mars and Jupiter. In his footnote Titius declared, "But should the Lord Architect have left that space empty? Not at all."[6] In 1768, the astronomer Johann Elert Bode made note of Titius's relationship in his *Anleitung zur Kenntniss des gestirnten Himmels* (English: *Instruction for the Knowledge of the Starry Heavens*) but did not credit Titius until later editions. It became known as "Bode's law".[7] When William Herschel discovered Uranus in 1781, the planet's orbit matched the law almost perfectly, leading astronomers to conclude that there had to be a planet between the orbits of Mars and Jupiter.

In 1800 the astronomer Baron Franz Xaver von Zach recruited 24 of his fellows into a club, the Vereinigte Astronomische Gesellschaft ("United Astronomical Society") which he informally dubbed the "Lilienthal Society"[8] for its meetings in Lilienthal, a small city near Bremen. Determined to bring the Solar System to order, the group became known as the "Himmelspolizei", or Celestial Police. Notable members included Herschel, the British Astronomer Royal Nevil Maskelyne, Charles Messier, and Heinrich Olbers.[9] The Society assigned to each astronomer a 15° region of the zodiac to search for the missing planet.[10]

Only a few months later, a non-member of the Celestial Police confirmed their expectations. On January 1, 1801, Giuseppe Piazzi, Chair of Astronomy at the University of Palermo, Sicily, found a tiny moving object in an orbit with exactly the radius predicted by the Titius–Bode law. He dubbed it Ceres, after the Roman goddess of the harvest and patron of Sicily. Piazzi initially believed it a comet, but its lack of a coma suggested it was a planet.[9] Fifteen months later, Heinrich Wilhelm Olbers discovered a second object in the same region, Pallas. Unlike the other known planets, the objects remained points of light even under the highest telescope magnifications, rather than resolving into discs. Apart from their rapid movement, they appeared indistinguishable from stars. Accordingly, in 1802 William Herschel suggested they be placed into a separate category, named *asteroids*, after the Greek *asteroeides*, meaning "star-like".[11] [12] Upon completing a series of observations of Ceres and Pallas, he concluded,[13]

> Neither the appellation of planets, nor that of comets, can with any propriety of language be given to these two stars ... They resemble small stars so much as hardly to be distinguished from them. From this, their asteroidal appearance, if I take my name, and call them Asteroids; reserving for myself however the liberty of changing that name, if another, more expressive of their nature, should occur.

Despite Herschel's coinage, for several decades it remained common practice to refer to these objects as planets.[5] By 1807, further investigation revealed two new objects in the region: 3 Juno and 4 Vesta.[14] The burning of Lilienthal in the Napoleonic wars brought this first period of discovery to a close,[14] and only in 1845 did astronomers detect another object (5 Astraea). Shortly thereafter new objects were found at an accelerating rate, and counting them among the planets became increasingly cumbersome. Eventually, they were dropped from the planet list as first suggested by Alexander von Humboldt in the early 1850s, and William Herschel's choice of nomenclature, "asteroids", gradually came into common use.[5]

The discovery of Neptune in 1846 led to the discrediting of the Titius–Bode law in the eyes of scientists, as its orbit was nowhere near the predicted position. To date, there is no scientific explanation for the law, and astronomers' consensus regards it as a coincidence.[15]

The expression "asteroid belt" came into use in the very early 1850s, although it is hard to pinpoint who coined the term. The first English use seems to be in the 1850 translation (by E. C. Otté) of Alexander von Humboldt's *Cosmos:*[16] "[...] and the regular appearance, about the 13th of November and the 11th of August, of shooting stars, which probably form part of a belt of asteroids intersecting the Earth's orbit and moving with planetary velocity". Other early appearances occur in Robert James Mann's *A Guide to the Knowledge of the Heavens,*[17] "The orbits of the asteroids are placed in a wide belt of space, extending between the extremes of [...]". The American astronomer Benjamin Peirce seems to have adopted that terminology and to have been one of its promoters.[18] One hundred asteroids had been located by mid-1868, and in 1891 the introduction of astrophotography by Max Wolf accelerated the rate of discovery still further.[19] A total of 1,000 asteroids had been found by 1921,[20] 10,000 by 1981,[21] and 100,000 by 2000.[22] Modern asteroid survey systems now use automated means to locate new minor planets in ever-increasing quantities.

Origin

Formation

In 1802, shortly after discovering Pallas, Heinrich Olbers suggested to William Herschel that Ceres and Pallas were fragments of a much larger planet that once occupied the Mars–Jupiter region, this planet having suffered an internal explosion or a cometary impact many million years before.[23] Over time, however, this hypothesis has fallen from favor. The large amount of energy that would have been required to destroy a planet, combined with the belt's low combined mass, which is only about 4% of the mass of the Earth's Moon,[1] do not support the hypothesis. Further, the significant chemical differences between the asteroids are difficult to explain if they come from the same planet.[24] Today, most scientists accept that, rather than fragmenting from a progenitor planet, the asteroids never formed a planet at all.

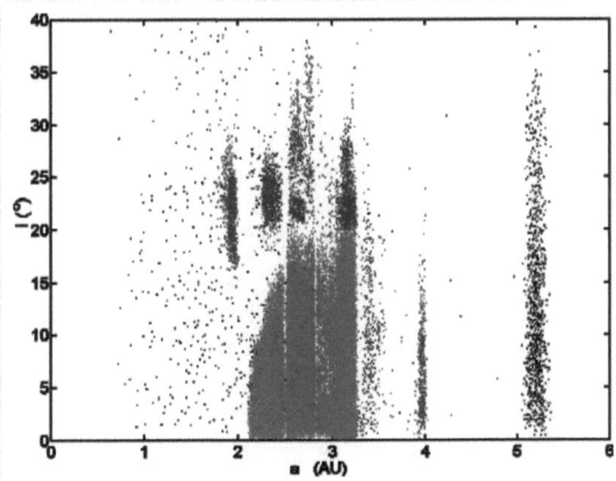

The asteroid belt showing the orbital inclinations versus distances from the Sun, with asteroids in the core region of the asteroid belt in red and other asteroids in blue

In general in the Solar System, planetary formation is thought to have occurred via a process comparable to the long-standing nebular hypothesis: a cloud of interstellar dust and gas collapsed under the influence of gravity to form a rotating disk of material that then further condensed to form the Sun and planets.[25] During the first few million years of the Solar System's history, an accretion process of sticky collisions caused the clumping of small particles, which gradually increased in size. Once the clumps reached sufficient mass, they could draw in other bodies through gravitational attraction and become planetesimals. This gravitational accretion led to the formation of the rocky planets and the gas giants.

Planetesimals within the region which would become the asteroid belt were too strongly perturbed by Jupiter's gravity to form a planet. Instead they continued to orbit the Sun as before, while occasionally colliding.[26] In regions where the average velocity of the collisions was too high, the shattering of planetesimals tended to dominate over accretion,[27] preventing the formation of planet-sized bodies. Orbital resonances occurred where the orbital period of an object in the belt formed an integer fraction of the orbital period of Jupiter, perturbing the object into a different orbit; the region lying between the orbits of Mars and Jupiter contains many such orbital resonances. As Jupiter migrated inward following its formation, these resonances would have swept across the asteroid belt, dynamically exciting the region's population and increasing their velocities relative to each other.[28]

During the early history of the Solar System, the asteroids melted to some degree, allowing elements within them to be partially or completely differentiated by mass. Some of the progenitor bodies may even have undergone periods of explosive volcanism and formed magma oceans. However, because of the relatively small size of the bodies, the period of melting was necessarily brief (compared to the much larger planets), and had generally ended about 4.5 billion years ago, in the first tens of millions of years of formation.[29] In August 2007, a study of zircon crystals in an Antarctic meteorite believed to have originated from 4 Vesta suggested that it, and by extension the rest of the asteroid belt, had formed rather quickly, within ten million years of the Solar System's origin.[30]

Evolution

The asteroids are not samples of the primordial Solar System. They have undergone considerable evolution since their formation, including internal heating (in the first few tens of millions of years), surface melting from impacts, space weathering from radiation, and bombardment by micrometeorites.[31] While some scientists refer to the asteroids as residual planetesimals,[32] other scientists consider them distinct.[33]

The current asteroid belt is believed to contain only a small fraction of the mass of the primordial belt. Computer simulations suggest that the original asteroid belt may have contained mass equivalent to the Earth. Primarily because of gravitational perturbations, most of the material was ejected from the belt within about a million years of formation, leaving behind less than 0.1% of the original mass.[26] Since their formation, the size distribution of the asteroid belt has remained relatively stable: there has been no significant increase or decrease in the typical dimensions of the main-belt asteroids.[34]

The 4:1 orbital resonance with Jupiter, at a radius 2.06 AU, can be considered the inner boundary of the asteroid belt. Perturbations by Jupiter send bodies straying there into unstable orbits. Most bodies formed inside the radius of this gap were swept up by Mars (which has an aphelion at 1.67 AU) or ejected by its gravitational perturbations in the early history of the Solar System.[35] The Hungaria asteroids lie closer to the Sun than the 4:1 resonance, but are protected from disruption by their high inclination.[36]

When the asteroid belt was first formed, the temperatures at a distance of 2.7 AU from the Sun formed a "snow line" below the freezing point of water. Planetesimals formed beyond this radius were able to accumulate ice.[37] [38] In 2006 it was announced that a population of comets had been discovered within the asteroid belt beyond the snow line, which may have provided a source of water for Earth's oceans. According to some models, there was insufficient outgassing of water during the Earth's formative period to form the oceans, requiring an external source such as a cometary bombardment.[39]

Characteristics

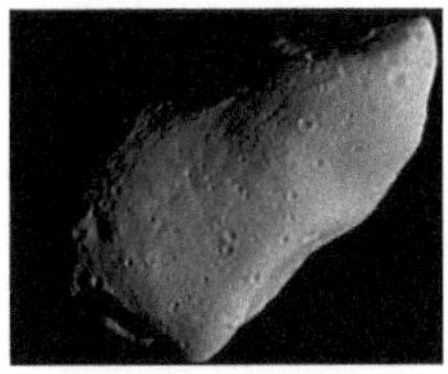

951 Gaspra, the first asteroid imaged by a spacecraft, as viewed during *Galileo*'s 1991 flyby; colors are exaggerated

Fragment of the Allende meteorite, a carbonaceous chondrite that fell to Earth in Mexico in 1969

Contrary to popular imagery, the asteroid belt is mostly empty. The asteroids are spread over such a large volume that it would be improbable to reach an asteroid without aiming carefully. Nonetheless, hundreds of thousands of asteroids are currently known, and the total number ranges in the millions or more, depending on the lower size cutoff. Over 200 asteroids are known to be larger than 100 km,[40] while a survey in the infrared wavelengths shows that the asteroid belt has 700,000 to 1.7 million asteroids with a diameter of 1 km or more.[41] The apparent magnitudes of most of the known asteroids are 11–19, with the median at about 16.[42]

The total mass of the asteroid belt is estimated to be 2.8×10^{21} to 3.2×10^{21} kilograms, which is just 4% of the mass of the Moon.[2] The four largest objects, Ceres, 4 Vesta, 2 Pallas, and 10 Hygiea, account for half of the belt's total mass, with almost one-third accounted for by Ceres alone.[3] [4]

Composition

The current belt consists primarily of three categories of asteroids: C-type or carbonaceous asteroids, S-type or silicate asteroids, and M-type or metallic asteroids.

Carbonaceous asteroids, as their name suggests, are carbon-rich and dominate the belt's outer regions.[43] Together they comprise over 75% of the visible asteroids. They are more red in hue than the other asteroids and have a very low albedo. Their surface composition is similar to carbonaceous chondrite meteorites. Chemically, their spectra match the primordial composition of the early Solar System, with only the lighter elements and volatiles removed.

S-type (silicate-rich) asteroids are more common toward the inner region of the belt, within 2.5 AU of the Sun.[43] [44] The spectra of their surfaces reveal the presence of silicates and some metal, but no significant carbonaceous compounds. This indicates that their materials have been significantly modified from their primordial composition, probably through melting and reformation. They have a relatively high albedo, and form about 17% of the total asteroid population.

M-type (metal-rich) asteroids form about 10% of the total population; their spectra resemble that of iron-nickel. Some are believed to have formed from the metallic cores of differentiated progenitor bodies that were disrupted through collision. However, there are also some silicate compounds that can produce a similar appearance. For example, the large M-type asteroid 22 Kalliope does not appear to be primarily composed of metal.[45] Within the asteroid belt, the number distribution of M-type asteroids peaks at a semi-major axis of about 2.7 AU.[46] It is not yet clear whether all M-types are compositionally similar, or whether it is a label for several varieties which do not fit neatly into the main C and S classes.[47]

One mystery of the asteroid belt is the relative rarity of V-type, or basaltic asteroids.[48] Theories of asteroid formation predict that objects the size of Vesta or larger should form crusts and mantles, which would be composed mainly of basaltic rock, resulting in more than half of all asteroids being composed either of basalt or olivine. Observations, however, suggest that 99 percent of the predicted basaltic material is missing.[49] Until 2001, most basaltic bodies discovered in the asteroid belt were believed to originate from the asteroid Vesta (hence their name V-type). However, the discovery of the asteroid 1459 Magnya revealed a slightly different chemical composition from the other basaltic asteroids discovered until then, suggesting a different origin.[49] This hypothesis was reinforced by the further discovery in 2007 of two asteroids in the outer belt, 7472 Kumakiri and (10537) 1991 RY_{16}, with differing basaltic composition that could not have originated from Vesta. These latter two are the only V-type asteroids discovered in the outer belt to date.[48]

The temperature of the asteroid belt varies with the distance from the Sun. For dust particles within the belt, typical temperatures range from 200 K (−73 °C) at 2.2 AU down to 165 K (−108 °C) at 3.2 AU[50] However, due to rotation, the surface temperature of an asteroid can vary considerably as the sides are alternately exposed to solar radiation and then to the stellar background.

Main-belt comets

Several otherwise unremarkable bodies in the outer belt show cometary activity. Since their orbits cannot be explained through capture of classical comets, it is thought that many of the outer asteroids may be icy, with the ice occasionally exposed to sublimation through small impacts. Main-belt comets may have been a major source of the Earth's oceans, since the deuterium-hydrogen ratio is too low for classical comets to have been the principal source.[51]

Orbits

Most asteroids within the asteroid belt have orbital eccentricities of less than 0.4, and an inclination of less than 30°. The orbital distribution of the asteroids reaches a maximum at an eccentricity of around 0.07 and an inclination below 4°.[42] Thus while a typical asteroid has a relatively circular orbit and lies near the plane of the ecliptic, some asteroid orbits can be highly eccentric or travel well outside the ecliptic plane.

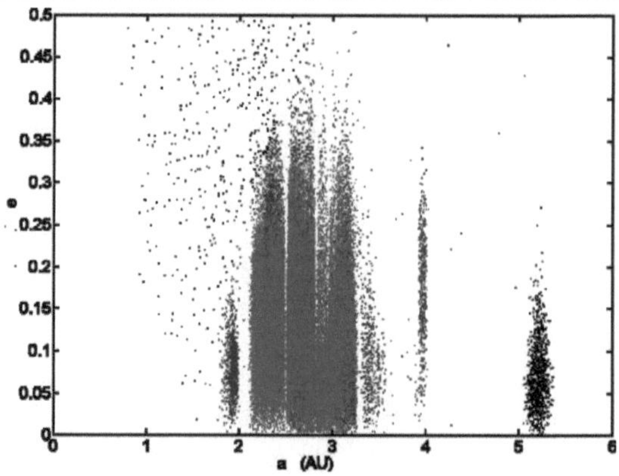

The asteroid belt (showing eccentricities), with the asteroid belt in red and blue ("core" region in red)

Sometimes, the term *main belt* is used to refer only to the more compact "core" region where the greatest concentration of bodies is found. This lies between the strong 4:1 and 2:1 Kirkwood gaps at 2.06 and 3.27 AU, and at orbital eccentricities less than roughly 0.33, along with orbital inclinations below about 20°. This "core" region contains approximately 93.4% of all numbered minor planets within the Solar System.[52]

Kirkwood gaps

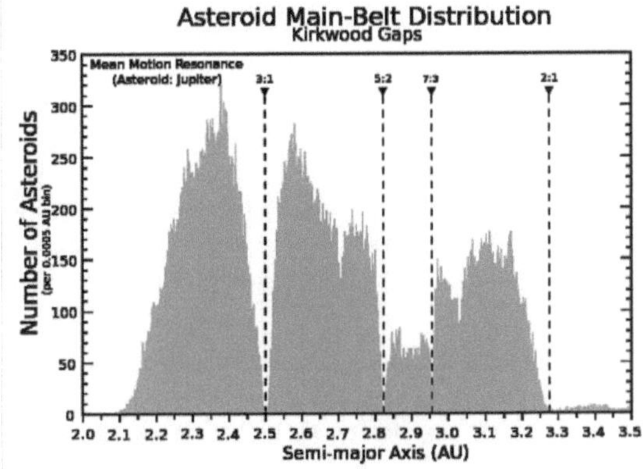

This chart shows the distribution of asteroid semi-major axes in the "core" of the asteroid belt. Black arrows point to the Kirkwood gaps, where orbital resonances with Jupiter destabilize orbits.

The semi-major axis of an asteroid is used to describe the dimensions of its orbit around the Sun, and its value determines the minor planet's orbital period. In 1866, Daniel Kirkwood announced the discovery of gaps in the distances of these bodies' orbits from the Sun. They were located at positions where their period of revolution about the Sun was an integer fraction of Jupiter's orbital period. Kirkwood proposed that the gravitational perturbations of the planet led to the removal of asteroids from these orbits.[53]

When the mean orbital period of an asteroid is an integer fraction of the orbital period of Jupiter, a mean-motion resonance with the gas giant is created that is sufficient to perturb an asteroid to new orbital elements. Asteroids that become located in the gap orbits (either primordially because of the migration of Jupiter's orbit,[54] or due to prior perturbations or collisions) are gradually nudged into different, random orbits with a larger or smaller semi-major axis.

The gaps are not seen in a simple snapshot of the locations of the asteroids at any one time because asteroid orbits are elliptical, and many asteroids still cross through the radii corresponding to the gaps. The actual spatial density of asteroids in these gaps does not differ significantly from the neighboring regions.[55]

The main gaps occur at the 3:1, 5:2, 7:3, and 2:1 mean-motion resonances with Jupiter. An asteroid in the 3:1 Kirkwood gap would orbit the Sun three times for each Jovian orbit, for instance. Weaker resonances occur at other semi-major axis values, with fewer asteroids found than nearby. (For example, an 8:3 resonance for asteroids with a semi-major axis of 2.71 AU.)[56]

The main or core population of the asteroid belt is sometimes divided into three zones, based on the most prominent Kirkwood gaps. Zone I lies between the 4:1 resonance (2.06 AU) and 3:1 resonance (2.5 AU) Kirkwood gaps. Zone II continues from the end of Zone I out to the 5:2 resonance gap (2.82 AU). Zone III extends from the outer edge of Zone II to the 2:1 resonance gap (3.28 AU).[57]

The asteroid belt may also be divided into the inner and outer belts, with the inner belt formed by asteroids orbiting nearer to Mars than the 3:1 Kirkwood gap (2.5 AU), and the outer belt formed by those asteroids closer to Jupiter's orbit. (Some authors subdivide the inner and outer belts at the 2:1 resonance gap (3.3 AU), while others suggest inner, middle, and outer belts.)

Collisions

The high population of the asteroid belt makes for a very active environment, where collisions between asteroids occur frequently (on astronomical time scales). Collisions between main-belt bodies with a mean radius of 10 km are expected to occur about once every 10 million years.[58] A collision may fragment an asteroid into numerous smaller pieces (leading to the formation of a new asteroid family). Conversely, collisions that occur at low relative speeds may also join two asteroids. After more than 4 billion years of such processes, the members of the asteroid belt now bear little resemblance to the original population.

The zodiacal light, created in part by dust from collisions in the asteroid belt

Along with the asteroid bodies, the asteroid belt also contains bands of dust with particle radii of up to a few hundred micrometres. This fine material is produced, at least in part, from collisions between asteroids, and by the impact of micrometeorites upon the asteroids. Due to the Poynting-Robertson effect, the pressure of solar radiation causes this dust to slowly spiral inward toward the Sun.[59]

The combination of this fine asteroid dust, as well as ejected cometary material, produces the zodiacal light. This faint auroral glow can be viewed at night extending from the direction of the Sun along the plane of the ecliptic. Particles that produce the visible zodiacal light average about 40 μm in radius. The typical lifetimes of such particles are about 700,000 years. Thus, to maintain the bands of dust, new particles must be steadily produced within the asteroid belt.[59]

Meteorites

Some of the debris from collisions can form meteoroids that enter the Earth's atmosphere.[60] Of the 50,000 meteorites found on Earth to date, 99.8 percent are believed to have originated in the asteroid belt.[61] A September 2007 study has suggested that a large-body collision undergone by the asteroid 298 Baptistina sent a number of fragments into the inner Solar System. The impacts of these fragments are believed to have created both Tycho crater on the Moon and Chicxulub crater in Mexico, the relict of the massive impact which is believed to have triggered the extinction of the dinosaurs 65 million years ago.[62]

Families and groups

In 1918, the Japanese astronomer Kiyotsugu Hirayama noticed that the orbits of some of the asteroids had similar parameters, forming families or groups.[63]

Approximately one-third of the asteroids in the asteroid belt are members of an asteroid family. These share similar orbital elements, such as semi-major axis, eccentricity, and orbital inclination as well as similar spectral features, all of which indicate a common origin in the breakup of a larger body. Graphical displays of these elements, for members of the asteroid belt, show concentrations indicating the presence of an asteroid family. There are about 20–30 associations that are almost certainly asteroid families. Additional groupings have been found that are less certain. Asteroid families can be confirmed when the members display common spectral features.[64] Smaller associations of asteroids are called groups or clusters.

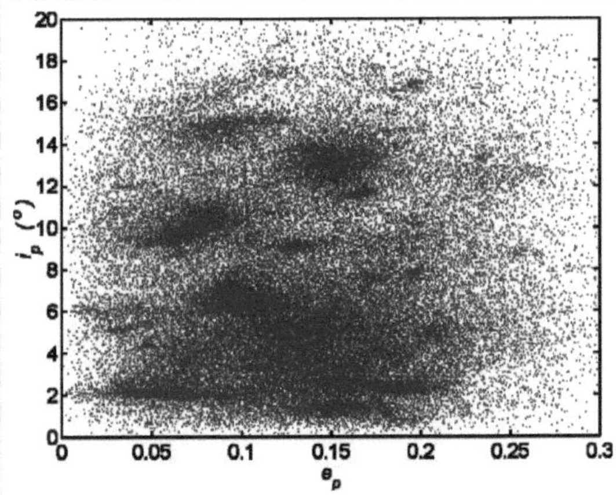

This plot of orbital inclination (i_p) versus eccentricity (e_p) for the numbered main-belt asteroids clearly shows clumpings representing asteroid families.

Some of the most prominent families in the asteroid belt (in order of increasing semi-major axes) are the Flora, Eunoma, Koronis, Eos, and Themis families.[46] The Flora family, one of the largest with more than 800 known members, may have formed from a collision less than a billion years ago.[65] The largest asteroid to be a true member of a family (as opposed to an interloper in the case of Ceres with the Gefion family) is 4 Vesta. The Vesta family is believed to have formed as the result of a crater-forming impact on Vesta. Likewise, the HED meteorites may also have originated from Vesta as a result of this collision.[66]

Three prominent bands of dust have been found within the asteroid belt. These have similar orbital inclinations as the Eos, Koronis, and Themis asteroid families, and so are possibly associated with those groupings.[67]

Periphery

Skirting the inner edge of the belt (ranging between 1.78 and 2.0 AU, with a mean semi-major axis of 1.9 AU) is the Hungaria family of minor planets. They are named after the main member, 434 Hungaria; the group contains at least 52 named asteroids. The Hungaria group is separated from the main body by the 4:1 Kirkwood gap and their orbits have a high inclination. Some members belong to the Mars-crossing category of asteroids, and gravitational perturbations by Mars are likely a factor in reducing the total population of this group.[68]

Another high-inclination group in the inner part of the asteroid belt is the Phocaea family. These are composed primarily of S-type asteroids, whereas the neighboring Hungaria family includes some E-types.[69] The Phocaea family orbit between 2.25 and 2.5 AU from the Sun.

Skirting the outer edge of the asteroid belt is the Cybele group, orbiting between 3.3 and 3.5 AU. These have a 7:4 orbital resonance with Jupiter. The Hilda family orbit between 3.5 and 4.2 AU, and have relatively circular orbits and a stable 3:2 orbital resonance with Jupiter. There are few asteroids beyond 4.2 AU, until Jupiter's orbit. Here the two families of Trojan asteroids can be found, which, at least for objects larger than 1 km, are approximately as numerous as the asteroids of the asteroid belt.[70]

New families

Some asteroid families have formed recently, in astronomical terms. The Karin Cluster apparently formed about 5.7 million years ago from a collision with a 33 km radius progenitor asteroid.[71] The Veritas family formed about 8.3 million years ago; evidence includes interplanetary dust recovered from ocean sediment.[72]

More recently, the Datura cluster appears to have formed about 450 thousand years ago from a collision with a main-belt asteroid. The age estimate is based on the probability of the members having their current orbits, rather than from any physical evidence. However, this cluster may have been a source for some zodiacal dust material.[73] Other recent cluster formations, such as the Iannini cluster (*circa* 1–5 million years ago), may have provided additional sources of this asteroid dust.[74]

Exploration

The first spacecraft to traverse the asteroid belt was Pioneer 10, which entered the region on July 16, 1972. At the time there was some concern that the debris in the belt would pose a hazard to the spacecraft, but it has since been safely traversed by 11 Earth-based craft without incident. Pioneer 11, Voyagers 1 and 2 and Ulysses passed through the belt without imaging any asteroids. Galileo imaged the asteroid 951 Gaspra in 1991 and 243 Ida in 1993, NEAR imaged 253 Mathilde in 1997, Cassini imaged 2685 Masursky in 2000, Stardust imaged 5535 Annefrank in 2002, New Horizons imaged 132524 APL in 2006, Rosetta imaged 2867 Šteins in 2008, and Dawn has been orbiting Vesta since July 2011.[75] Due to the low density of materials within the belt, the odds of a probe running into an asteroid are now estimated at less than one in a billion.[76]

Artist's concept of the Dawn spacecraft with Vesta (left) and Ceres (right)

Most belt asteroids imaged to date have come from brief flyby opportunities by probes headed for other targets. Only the Dawn, NEAR and Hayabusa missions have studied asteroids for a protracted period in orbit and at the surface. Dawn explored Vesta from July 2011 to September 2012, and is currently en route to Ceres for a 2015 rendezvous. If the probe is still operational after examining Ceres, an extended mission could allow additional exploration, possibly a flyby of Pallas.[77]

See also

- Asteroid mining
- Asteroids in astrology
- Asteroids in fiction
- Colonization of the asteroids
- Debris disk
- Kuiper belt

References

[1] Krasinsky, G. A.; Pitjeva, E. V.; Vasilyev, M. V.; Yagudina, and E. I. (July 2002). "Hidden Mass in the Asteroid Belt". *Icarus* **158** (1): 98–105. Bibcode 2002Icar..158...98K. doi:10.1006/icar.2002.6837.

[2] Pitjeva, E. V. (2005). "High-Precision Ephemerides of Planets—EPM and Determination of Some Astronomical Constants" (http://iau-comm4.jpl.nasa.gov/EPM2004.pdf) (PDF). *Solar System Research* **39** (3): 176. Bibcode 2005SoSyR..39..176P. doi:10.1007/s11208-005-0033-2. .

[3] For recent estimates of the masses of Ceres, 4 Vesta, 2 Pallas and 10 Hygiea, see the references in the infoboxes of their respective articles.

[4] Yeomans, Donald K. (July 13, 2006). "JPL Small-Body Database Browser" (http://ssd.jpl.nasa.gov/sbdb.cgi). NASA JPL. Archived (http://web.archive.org/web/20100929043420/http://ssd.jpl.nasa.gov/sbdb.cgi) from the original on 29 September 2010. . Retrieved 2010-09-27.

[5] Hilton, J. (2001). "When Did the Asteroids Become Minor Planets?" (http://www.usno.navy.mil/USNO/astronomical-applications/astronomical-information-center/minor-planets). *US Naval Observatory (USNO)*. . Retrieved 2007-10-01.

[6] "Dawn: A Journey to the Beginning of the Solar System" (http://www-ssc.igpp.ucla.edu/dawn/background.html). *Space Physics Center: UCLA*. 2005. . Retrieved 2007-11-03.

[7] Hoskin, Michael. "Bode's Law and the Discovery of Ceres" (http://www.astropa.unipa.it/HISTORY/hoskin.html). *Churchill College, Cambridge*. . Retrieved 2010-07-12.

[8] Linda T. Elkins-Tanton, *Asteroids, Meteorites, and Comets*, 2010:10

[9] "Call the police! The story behind the discovery of the asteroids". *Astronomy Now* (June 2007): 60–61.

[10] Pogge, Richard (2006). "An Introduction to Solar System Astronomy: Lecture 45: Is Pluto a Planet?" (http://www.astronomy.ohio-state.edu/~pogge/Ast161/Unit6/dwarfs.html). *An Introduction to Solar System Astronomy*. Ohio State University. . Retrieved 2007-11-11.

[11] Harper, Douglas (2010). "Asteroid" (http://www.etymonline.com/index.php?search=asteroid&searchmode=none). *Online Etymology Dictionary*. Etymology Online. . Retrieved 2011-04-15.

[12] DeForest, Jessica (2000). "Greek and Latin Roots" (http://www.msu.edu/~defores1/gre/roots/gre_rts_afx2.htm). Michigan State University. Archived (http://web.archive.org/web/20070812155525/http://www.msu.edu/~defores1/gre/roots/gre_rts_afx2.htm) from the original on 12 August 2007. . Retrieved 2007-07-25.

[13] Cunningham, Clifford (1984). "William Hershel and the First Two Asteroids". *Dance Hall Observatory, Ontario* **11**: 3. Bibcode 1984MPBu...11....3C.

[14] Staff (2002). "Astronomical Serendipity" (http://dawn.jpl.nasa.gov/DawnCommunity/flashbacks/fb_06.asp). NASA JPL. . Retrieved 2007-04-20.

[15] "Is it a coincidence that most of the planets fall within the Titius-Bode law's boundaries?" (http://www.astronomy.com/asy/default.aspx?c=a&id=4494). *astronomy.com*. . Retrieved 2007-10-16.

[16] von Humboldt, Alexander (1850). *Cosmos: A Sketch of a Physical Description of the Universe*. **1**. Harper & Brothers, New York (NY). p. 44. ISBN 0-8018-5503-9.

[17] Mann, Robert James (1852). *A Guide to the Knowledge of the Heavens*. Jarrold. p. 171. and 1853, p. 216

[18] "Further Investigation relative to the form, the magnitude, the mass, and the orbit of the Asteroid Planets" (http://books.google.com/?id=hhQAAAAAMAAJ&pg=PA191&dq=asteroid+belt). *The Edinburgh New Philosophical Journal* **5**: 191. January–April 1857. .: "[Professor Peirce] then observed that the analogy between the ring of Saturn and the belt of the asteroids was worthy of notice."

[19] Hughes, David W. (2007). "A Brief History of Asteroid Spotting" (http://www.open2.net/sciencetechnologynature/planetsbeyond/asteroids/history.html). BBC. . Retrieved 2007-04-20.

[20] Moore, Patrick; Rees, Robin (2011). *Patrick Moore's Data Book of Astronomy* (2nd ed.). Cambridge University Press. p. 156. ISBN 0-521-89935-4.

[21] Manley, Scott (August 25, 2010). *Asteroid Discovery from 1980 to 2010* (http://www.youtube.com/watch?v=S_d-gs0WoUw). . Retrieved 2011-04-15.

[22] "MPC Archive Statistics" (http://www.minorplanetcenter.org/iau/lists/ArchiveStatistics.html). IAU Minor Planet Center. . Retrieved 2011-04-04.

[23] "A Brief History of Asteroid Spotting" (http://www.open2.net/sciencetechnologynature/planetsbeyond/asteroids/history.html). *Open2.net*. . Retrieved 2007-05-15.

[24] Masetti, M.; and Mukai, K. (December 1, 2005). "Origin of the Asteroid Belt" (http://imagine.gsfc.nasa.gov/docs/ask_astro/answers/980810a.html). NASA Goddard Spaceflight Center. . Retrieved 2007-04-25.

[25] Watanabe, Susan (July 20, 2001). "Mysteries of the Solar Nebula" (http://www.jpl.nasa.gov/news/features.cfm?feature=520). NASA. . Retrieved 2007-04-02.

[26] Petit, J.-M.; Morbidelli, A.; and Chambers, J. (2001). "The Primordial Excitation and Clearing of the Asteroid Belt" (http://www.gps.caltech.edu/classes/ge133/reading/asteroids.pdf) (PDF). *Icarus* **153** (2): 338–347. Bibcode 2001Icar..153..338P. doi:10.1006/icar.2001.6702. Archived (http://web.archive.org/web/20070221085835/http://www.gps.caltech.edu/classes/ge133/reading/asteroids.pdf) from the original on 21 February 2007. . Retrieved 2007-03-22.

[27] Edgar, R.; and Artymowicz, P. (2004). "Pumping of a Planetesimal Disc by a Rapidly Migrating Planet" (http://www.astro.su.se/~pawel/edgar+artymowicz.pdf) (PDF). *Monthly Notices of the Royal Astronomical Society* **354** (3): 769–772. arXiv:astro-ph/0409017. Bibcode 2004MNRAS.354..769E. doi:10.1111/j.1365-2966.2004.08238.x. . Retrieved 2007-04-16.

[28] Scott, E. R. D. (March 13–17, 2006). "Constraints on Jupiter's Age and Formation Mechanism and the Nebula Lifetime from Chondrites and Asteroids" (http://adsabs.harvard.edu/abs/2006LPI....37.2367S). *Proceedings 37th Annual Lunar and Planetary Science Conference*. League City, Texas: Lunar and Planetary Society. . Retrieved 2007-04-16.

[29] Taylor, G. J.; Keil, K.; McCoy, T.; Haack, H.; and Scott, E. R. D.; Keil; McCoy; Haack; Scott (1993). "Asteroid differentiation – Pyroclastic volcanism to magma oceans". *Meteoritics* **28** (1): 34–52. Bibcode 1993Metic..28...34T.

[30] Kelly, Karen (2007). "U of T researchers discover clues to early solar system" (http://webapps.utsc.utoronto.ca/ose/story.php?id=665). *University of Toronto*. . Retrieved 2010-07-12.

[31] Clark, B. E.; Hapke, B.; Pieters, C.; Britt, D.; Hapke; Pieters; Britt (2002). "Asteroid Space Weathering and Regolith Evolution". *University of Arizona*: 585. Bibcode 2002aste.conf..585C. Gaffey, Michael J. (1996). "The Spectral and Physical Properties of Metal in Meteorite Assemblages: Implications for Asteroid Surface Materials". *Icarus (ISSN 0019-1035)* **66** (3): 468. Bibcode 1986Icar...66..468G. doi:10.1016/0019-1035(86)90086-2. Keil, K. (2000). "Thermal alteration of asteroids: evidence from meteorites" (http://www.ingentaconnect.com/content/els/00320633/2000/00000048/00000010/art00054). *Planetary and Space Science*. . Retrieved 2007-11-08. Baragiola, R. A.; Duke, C. A.; Loeffler, M.; McFadden, L. A.; and Sheffield, J.; Duke; Loeffler; McFadden; Sheffield (2003). "Impact of ions and micrometeorites on mineral surfaces: Reflectance changes and production of atmospheric species in airless solar system bodies". *EGS - AGU - EUG Joint Assembly*: 7709. Bibcode 2003EAEJA.....7709B.

[32] Chapman, C. R.; Williams, J. G.; Hartmann, W. K. (1978). "The asteroids". *Annual review of astronomy and astrophysics* **16**: 33–75. Bibcode 1978ARA&A..16...33C. doi:10.1146/annurev.aa.16.090178.000341.

[33] Kracher, A. (2005). "Asteroid 433 Eros and partially differentiated planetesimals: bulk depletion versus surface depletion of sulfur" (http://www.cosis.net/abstracts/EGU05/03788/EGU05-J-03788.pdf) (PDF). *Ames Laboratory*. Archived (http://web.archive.org/web/20071128200221/http://www.cosis.net/abstracts/EGU05/03788/EGU05-J-03788.pdf) from the original on 28 November 2007. . Retrieved 2007-11-08.

[34] Stiles, Lori (September 15, 2005). "Asteroids Caused the Early Inner Solar System Cataclysm" (http://uanews.org/cgi-bin/WebObjects/UANews.woa/7/wa/SRStoryDetails?ArticleID=11692). University of Arizona News. . Retrieved 2007-04-18.

[35] Alfvén, H.; Arrhenius, G. (1976). "The Small Bodies" (http://history.nasa.gov/SP-345/ch4.htm). *SP-345 Evolution of the Solar System*. NASA. Archived (http://web.archive.org/web/20070513081833/http://history.nasa.gov/SP-345/ch4.htm) from the original on 13 May 2007. . Retrieved 2007-04-12.

[36] Spratt, Christopher E. (April 1990). "The Hungaria group of minor planets". *Journal of the Royal Astronomical Society of Canada* **84**: 123–131. Bibcode 1990JRASC..84..123S.

[37] Lecar, M.; Podolak, M.; Sasselov, D.; Chiang, E. (2006). "Infrared cirrus – New components of the extended infrared emission". *The Astrophysical Journal* **640** (2): 1115–1118. arXiv:astro-ph/0602217. Bibcode 2006ApJ...640.1115L. doi:10.1086/500287.

[38] Berardelli, Phil (March 23, 2006). "Main-Belt Comets May Have Been Source Of Earths Water" (http://www.spacedaily.com/reports/Main_Belt_Comets_May_Have_Been_Source_Of_Earths_Water.html). Space Daily. Archived (http://web.archive.org/web/20071011171815/http://spacedaily.com/reports/Main_Belt_Comets_May_Have_Been_Source_Of_Earths_Water.html) from the original on 11 October 2007. . Retrieved 2007-10-27.

[39] Lakdawalla, Emily (April 28, 2006). "Discovery of a Whole New Type of Comet" (http://www.planetary.org/blog/article/00000551/). The Planetary Society. Archived (http://web.archive.org/web/20070501211319/http://www.planetary.org/blog/article/00000551/) from the original on 1 May 2007. . Retrieved 2007-04-20.

[40] Yeomans, Donald K. (April 26, 2007). "JPL Small-Body Database Search Engine" (http://ssd.jpl.nasa.gov/sbdb_query.cgi). NASA JPL. . Retrieved 2007-04-26. – search for asteroids in the main belt regions with a diameter >100.

[41] Tedesco, E. F.; and Desert, F.-X. (2002). "The Infrared Space Observatory Deep Asteroid Search". *The Astronomical Journal* **123** (4): 2070–2082. Bibcode 2002AJ....123.2070T. doi:10.1086/339482.

[42] Williams, Gareth (September 25, 2010). "Distribution of the Minor Planets" (http://www.minorplanetcenter.org/iau/lists/MPDistribution.html). Minor Planets Center. . Retrieved 2010-10-27.

[43] Wiegert, P.; Balam, D.; Moss, A.; Veillet, C.; Connors, M.; and Shelton, I. (2007). "Evidence for a Color Dependence in the Size Distribution of Main-Belt Asteroids" (http://astro.uwo.ca/~wiegert/papers/2007AJ.133.1609.pdf). *The Astronomical Journal* **133** (4): 1609–1614. arXiv:astro-ph/0611310. Bibcode 2007AJ....133.1609W. doi:10.1086/512128. . Retrieved 2008-09-06.

[44] Clark, B. E. (1996). "New News and the Competing Views of Asteroid Belt Geology". *Lunar and Planetary Science* **27**: 225–226. Bibcode 1996LPI....27..225C.

[45] Margot, J. L.; and Brown, M. E. (2003). "A Low-Density M-type Asteroid in the Main Belt". *Science* **300** (5627): 1939–1942. Bibcode 2003Sci...300.1939M. doi:10.1126/science.1085844. PMID 12817147.

[46] Lang, Kenneth R. (2003). "Asteroids and meteorites" (http://ase.tufts.edu/cosmos/print_images.asp?id=15). NASA's Cosmos. . Retrieved 2007-04-02.

[47] Mueller, M.; Harris, A. W.; Delbo, M.; and the MIRSI Team; Harris; Delbo (2005). "21 Lutetia and other M-types: Their sizes, albedos, and thermal properties from new IRTF measurements". *Bulletin of the American Astronomical Society* **37**: 627. Bibcode 2005DPS....37.0702M.

[48] Duffard, R. D.; Roig, F. (July 14–18, 2008). "Two New Basaltic Asteroids in the Main Belt?". *Asteroids, Comets, Meteors 2008*. Baltimore, Maryland. arXiv:0704.0230. Bibcode 2008LPICo1405.8154D.

[49] Than, Ker (2007). "Strange Asteroids Baffle Scientists" (http://www.space.com/scienceastronomy/070821_basalt_asteroid.html). *space.com*. . Retrieved 2007-10-14.

[50] Low, F. J.; *et al.* (1984). "Infrared cirrus – New components of the extended infrared emission". *Astrophysical Journal, Part 2 – Letters to the Editor* **278**: L19–L22. Bibcode 1984ApJ...278L..19L. doi:10.1086/184213.

[51] "Interview with David Jewitt" (http://www.youtube.com/watch?v=B1W4NTmI5Bk). Youtube.com. 2007-01-05. . Retrieved 2011-05-21.

[52] This value was obtained by a simple count up of all bodies in that region using data for 120437 numbered minor planets from the Minor Planet Center orbit database (http://www.minorplanetcenter.org/iau/MPCORB.html), dated February 8, 2006.

[53] Fernie, J. Donald (1999). "The American Kepler" (http://www.americanscientist.org/issues/pub/1999/9/the-american-kepler/2). *The Americal Scientist* **87** (5): 398. . Retrieved 2007-02-04.

[54] Liou, Jer-Chyi; and Malhotra, Renu (1997). "Depletion of the Outer Asteroid Belt" (http://www.sciencemag.org/cgi/content/full/275/5298/375). *Science* **275** (5298): 375–377. Bibcode 1997Sci...275..375L. doi:10.1126/science.275.5298.375. PMID 8994031. . Retrieved 2007-08-01.

[55] McBride, N.; and Hughes, D. W.; Hughes (1990). "The spatial density of asteroids and its variation with asteroidal mass". *Monthly Notices of the Royal Astronomical Society* **244**: 513–520. Bibcode 1990MNRAS.244..513M.

[56] Ferraz-Mello, S. (June 14–18, 1993). "Kirkwood Gaps and Resonant Groups" (http://adsabs.harvard.edu/abs/1994IAUS..160..175F). *proceedings of the 160th International Astronomical Union*. Belgirate, Italy: Kluwer Academic Publishers. pp. 175–188. . Retrieved 2007-03-28.

[57] Klacka, Jozef (1992). "Mass distribution in the asteroid belt". *Earth, Moon, and Planets* **56** (1): 47–52. Bibcode 1992EM&P...56...47K. doi:10.1007/BF00054599.

[58] Backman, D. E. (March 6, 1998). "Fluctuations in the General Zodiacal Cloud Density" (http://astrobiology.arc.nasa.gov/workshops/zodiac/backman/backman_toc.html). *Backman Report*. NASA Ames Research Center. . Retrieved 2007-04-04.

[59] Reach, William T. (1992). "Zodiacal emission. III – Dust near the asteroid belt". *Astrophysical Journal* **392** (1): 289–299. Bibcode 1992ApJ...392..289R. doi:10.1086/171428.

[60] Kingsley, Danny (May 1, 2003). "Mysterious meteorite dust mismatch solved" (http://abc.net.au/science/news/stories/s843594.htm). ABC Science. . Retrieved 2007-04-04.

[61] "Meteors and Meteorites" (http://www.nasa.gov/pdf/145945main_Meteors.Meteorites.Lithograph.pdf). NASA. . Retrieved 2012-01-12.

[62] "Breakup event in the main asteroid belt likely caused dinosaur extinction 65 million years ago" (http://www.physorg.com/news108218928.html). *Southwest Research Institute*. 2007. Archived (http://web.archive.org/web/20071011151507/http://physorg.com/news108218928.html) from the original on 11 October 2007. . Retrieved 2007-10-14.

[63] Hughes, David W. (2007). "Finding Asteroids In Space" (http://www.open2.net/sciencetechnologynature/planetsbeyond/asteroids/finding.html). BBC. . Retrieved 2007-04-20.

[64] Lemaitre, Anne (August 31-September 4, 2004). "Asteroid family classification from very large catalogues" (http://adsabs.harvard.edu/abs/2005dpps.conf..135L). *Proceedings Dynamics of Populations of Planetary Systems*. Belgrade, Serbia and Montenegro: Cambridge University Press. pp. 135–144. . Retrieved 2007-04-15.

[65] Martel, Linda M. V. (March 9, 2004). "Tiny Traces of a Big Asteroid Breakup" (http://www.psrd.hawaii.edu/Mar04/fossilMeteorites.html). Planetary Science Research Discoveries. Archived (http://web.archive.org/web/20070401115157/http://www.psrd.hawaii.edu/Mar04/fossilMeteorites.html) from the original on 1 April 2007. . Retrieved 2007-04-02.

[66] Drake, Michael J. (2001). "The eucrite/Vesta story". *Meteoritics & Planetary Science* **36** (4): 501–513. Bibcode 2001M&PS...36..501D. doi:10.1111/j.1945-5100.2001.tb01892.x.

[67] Love, S. G.; and Brownlee, D. E. (1992). "The IRAS dust band contribution to the interplanetary dust complex – Evidence seen at 60 and 100 microns". *Astronomical Journal* **104** (6): 2236–2242. Bibcode 1992AJ....104.2236L. doi:10.1086/116399.

[68] Spratt, Christopher E. (1990). "The Hungaria group of minor planets". *Journal of the Royal Astronomical Society of Canada* **84** (2): 123–131. Bibcode 1990JRASC..84..123S.

[69] Carvano, J. M.; Lazzaro, D.; Mothé-Diniz, T.; Angeli, C. A.; and Florczak, M. (2001). "Spectroscopic Survey of the Hungaria and Phocaea Dynamical Groups". *Icarus* **149** (1): 173–189. Bibcode 2001Icar..149..173C. doi:10.1006/icar.2000.6512.

[70] Dymock, Roger (2010). *Asteroids and Dwarf Planets and How to Observe Them* (http://books.google.com/books?id=vQcAnwt_87sC&pg=PA24). Astronomers' Observing Guides. Springer. p. 24. ISBN 1-4419-6438-X. . Retrieved 2011-04-04.

[71] Nesvorný, David; *et al.* (August 2006). "Karin cluster formation by asteroid impact". *Icarus* **183** (2): 296–311. Bibcode 2006Icar..183..296N. doi:10.1016/j.icarus.2006.03.008.

[72] McKee, Maggie (January 18, 2006). "Eon of dust storms traced to asteroid smash" (http://space.newscientist.com/channel/solar-system/comets-asteroids/dn8603). New Scientist Space. . Retrieved 2007-04-15.

[73] Nesvorný, D.; Vokrouhlick, D.; and Bottke, W. F. (2006). "The Breakup of a Main-Belt Asteroid 450 Thousand Years Ago" (http://www.sciencemag.org/cgi/content/full/312/5779/1490). *Science* **312** (5779): 1490. Bibcode 2006Sci...312.1490N. doi:10.1126/science.1126175. PMID 16763141. . Retrieved 2007-04-15.

[74] Nesvorný, D.; Bottke, W. F.; Levison, H. F.; and Dones, L. (2003). "Recent Origin of the Solar System Dust Bands" (http://iopscience.iop.org/0004-637X/591/1/486/pdf/0004-637X_591_1_486.pdf). *The Astrophysical Journal* **591** (1): 486–497. Bibcode 2003ApJ...591..486N. doi:10.1086/374807. . Retrieved 2007-04-15.

[75] Barucci, M. A.; Fulchignoni, M.; and Rossi, A. (2007). "Rosetta Asteroid Targets: 2867 Steins and 21 Lutetia". *Space Science Reviews* **128** (1–4): 67–78. Bibcode 2007SSRv..128...67B. doi:10.1007/s11214-006-9029-6.

[76] Stern, Alan (June 2, 2006). "New Horizons Crosses The Asteroid Belt" (http://www.spacedaily.com/reports/New_Horizons_Crosses_The_Asteroid_Belt.html). Space Daily. . Retrieved 2007-04-14.

[77] Staff (April 10, 2007). "Dawn Mission Home Page" (http://dawn.jpl.nasa.gov/). NASA JPL. Archived (http://web.archive.org/web/20070411053350/http://dawn.jpl.nasa.gov/) from the original on 11 April 2007. . Retrieved 2007-04-14.

Further reading

- Elkins-Tanton, Linda T. (2006). *Asteroids, Meteorites, and Comets* (First ed.). New York: Chelsea House. ISBN 0-8160-5195-X.

External links

- Asteroid Discovery from 1980 to 2010 (http://www.youtube.com/watch?v=S_d-gs0WoUw)
- Arnett, William A. (February 26, 2006). "Asteroids" (http://www.nineplanets.org/asteroids.html). The Nine Planets. Archived (http://web.archive.org/web/20070418220152/http://www.nineplanets.org/asteroids.html) from the original on 18 April 2007. Retrieved 2007-04-20.
- Asteroids Page (http://solarsystem.nasa.gov/planets/profile.cfm?Object=Asteroids) at NASA's Solar System Exploration (http://solarsystem.nasa.gov)
- Cain, Fraser. "The Asteroid Belt" (http://www.astronomycast.com/astronomy/episode-55-the-asteroid-belt/). Universe Today. Archived (http://web.archive.org/web/20080307033324/http://www.astronomycast.com/astronomy/episode-55-the-asteroid-belt/) from the original on 7 March 2008. Retrieved 2008-04-01.
- Hsieh, Henry H. (March 1, 2006). "Main-Belt Comets" (http://star.pst.qub.ac.uk/~hhh/mbcs.shtml). University of Hawaii. Retrieved 2007-04-20.
- "Main Asteroid Belt" (http://www.solstation.com/stars/asteroid.htm). Sol Company. Archived (http://web.archive.org/web/20070515163455/http://www.solstation.com/stars/asteroid.htm) from the original on 15 May 2007. Retrieved 2007-04-20.
- Munsell, Kirk (September 16, 2005). "Asteroids: Overview" (http://solarsystem.nasa.gov/planets/profile.cfm?Object=Asteroids). NASA's Solar System Exploration (http://solarsystem.nasa.gov). Archived (http://web.archive.org/web/20070524185042/http://solarsystem.nasa.gov/planets/profile.cfm?Object=Asteroids) from the original on 24 May 2007. Retrieved 2007-05-26.
- Plots of eccentricity vs. semi-major axis (http://burro.astr.cwru.edu/stu/media/asteroid_all_axisvecc.jpg) and inclination vs. semi-major axis (http://burro.astr.cwru.edu/stu/media/asteroid_all_axisvincl.jpg) at Asteroid Dynamic Site
- Staff (October 31, 2006). "Asteroids" (http://nssdc.gsfc.nasa.gov/planetary/planets/asteroidpage.html). NASA. Archived (http://web.archive.org/web/20070411201013/http://nssdc.gsfc.nasa.gov/planetary/planets/asteroidpage.html) from the original on 11 April 2007. Retrieved 2007-04-20.
- Staff (2007). "Space Topics: Asteroids and Comets" (http://www.planetary.org/explore/topics/asteroids_and_comets/facts.html). The Planetary Society. Archived (http://web.archive.org/web/20070428143402/http://www.planetary.org/explore/topics/asteroids_and_comets/facts.html) from the original on 28 April 2007. Retrieved 2007-04-20.

Asteroid

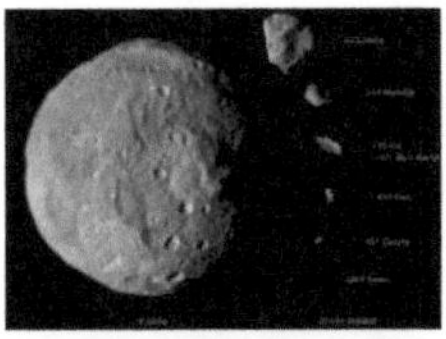

A composite image, to scale, of the asteroids which have been imaged at high resolution. As of 2011 they are, from largest to smallest: 4 Vesta, 21 Lutetia, 253 Mathilde, 243 Ida and its moon Dactyl, 433 Eros, 951 Gaspra, 2867 Šteins, 25143 Itokawa.

The largest asteroid in the previous image, Vesta (left), with Ceres (center) and Earth's Moon (right) shown to scale.

Asteroids are a class of small Solar System bodies in orbit around the Sun. They have also been called **planetoids**, especially the larger ones. These terms have historically been applied to any astronomical object orbiting the Sun that did not show the disk of a planet and was not observed to have the characteristics of an active comet, but as small objects in the outer Solar System were discovered, their volatile-based surfaces were found to more closely resemble comets, and so were often distinguished from traditional asteroids.[1] Thus the term *asteroid* has come increasingly to refer specifically to the small bodies of the inner Solar System out to the orbit of Jupiter, which are usually rocky or metallic. They are grouped with the outer bodies—centaurs, Neptune trojans, and trans-Neptunian objects—as minor planets, which is the term preferred in astronomical circles.[2] This article will restrict the use of the term "asteroid" to the minor planets of the inner Solar System.

There are millions of asteroids, many thought to be the shattered remnants of planetesimals, bodies within the young Sun's solar nebula that never grew large enough to become planets.[3] A large majority of known asteroids orbit in the asteroid belt between the orbits of Mars and Jupiter or co-orbital with Jupiter (the Jupiter Trojans). However, other orbital families exist with significant populations, including the near-Earth asteroids. Individual asteroids are classified by their characteristic spectra, with the majority falling into three main groups: C-type, S-type, and M-type. These were named after and are generally identified with carbon-rich, stony, and metallic compositions, respectively.

Naming

A newly discovered asteroid is given a provisional designation (such as 2002 AT_4) consisting of the year of discovery and an alphanumeric code indicating the half-month of discovery and the sequence within that half-month. Once an asteroid's orbit has been confirmed, it is given a number, and later may also be given a name (e.g. 433 Eros). The formal naming convention uses parentheses around the number (e.g. (433) Eros), but dropping the parentheses is quite common. Informally, it is common to drop the number altogether, or to drop it after the first mention when a name is repeated in running text.

Symbols

The first asteroids to be discovered were assigned iconic symbols like the ones traditionally used to designate the planets. By 1855 there were two dozen asteroid symbols, which often occurred in multiple variants.[4]

Asteroid	Symbol		Year
Ceres		Ceres' scythe, reversed to double as the letter *C*	1801
2 Pallas		Athena's (Pallas') spear	1801
3 Juno		A star mounted on a scepter, for Juno, the Queen of Heaven	1804
4 Vesta		The altar and sacred fire of Vesta	1807
5 Astraea		A scale, or an inverted anchor, symbols of justice	1845
6 Hebe		Hebe's cup	1847
7 Iris		A rainbow (*iris*) and a star	1847
8 Flora		A flower (*flora*) (spec. the Rose of England)	1847
9 Metis		The eye of wisdom and a star	1848
10 Hygiea		Hygiea's serpent and a star, or the Rod of Asclepius	1849
11 Parthenope		A harp, or a fish and a star; symbols of the sirens	1850
12 Victoria		The laurels of victory and a star	1850
13 Egeria		A shield, symbol of Egeria's protection, and a star	1850
14 Irene		A dove carrying an olive-branch (symbol of *irene* 'peace') with a star on its head,[5] or an olive branch, a flag of truce, and a star	1851
15 Eunomia		A heart, symbol of good order (*eunomia*), and a star	1851
16 Psyche		A butterfly's wing, symbol of the soul (*psyche*), and a star	1852
17 Thetis		A dolphin, symbol of Thetis, and a star	1852
18 Melpomene		The dagger of Melpomene, and a star	1852
19 Fortuna		The wheel of fortune and a star	1852
26 Proserpina		Proserpina's pomegranate	1853
28 Bellona		Bellona's whip and lance[6]	1854
29 Amphitrite		The shell of Amphitrite and a star	1854
35 Leukothea		A lighthouse beacon, symbol of Leucothea[7]	1855
37 Fides		The cross of faith (*fides*)[8]	1855

In 1851,[9] after the fifteenth asteroid (Eunomia) had been discovered, Johann Franz Encke made a major change in the upcoming 1854 edition of the *Berliner Astronomisches Jahrbuch* (BAJ, *Berlin Astronomical Yearbook*). He introduced a disk (circle), a traditional symbol for a star, as the generic symbol for an asteroid. The circle was then numbered in order of discovery to indicate a specific asteroid (although he assigned ① to the fifth, Astraea, while continuing to designate the first four only with their existing iconic symbols). The numbered-circle convention was quickly adopted by the astronomical community, and the next asteroid to be discovered (16 Psyche, in 1852) was the first to be designated in this manner at the time of its discovery. However, Psyche was also given an iconic symbol, as were a few other asteroids discovered over the next few years (see chart above). 20 Massalia was the first asteroid that was not assigned a symbol, and no additional iconic symbols were created after the 1855 discovery of 37 Fides.[10] That year Astraea's number was bumped up to ⑤, but Ceres through Vesta would not be listed by their numbers until the 1867 edition. The circle would become a pair of parentheses, and the parentheses sometimes omitted altogether over the next few decades, leading to the modern convention.[5]

Discovery

The first asteroid to be discovered, Ceres, was found in 1801 by Giuseppe Piazzi, and was originally considered to be a new planet.[11] This was followed by the discovery of other similar bodies, which, with the equipment of the time, appeared to be points of light, like stars, showing little or no planetary disc, though readily distinguishable from stars due to their apparent motions. This prompted the astronomer Sir William Herschel to propose the term "asteroid", coined in Greek as αστεροειδής *asteroeidēs* 'star-like, star-shaped', from ancient Greek ἀστήρ *astēr* 'star, planet'. In the early second half of the nineteenth century, the terms "asteroid" and "planet" (not always qualified as "minor") were still used interchangeably; for example, the *Annual of Scientific Discovery for 1871* [12], page 316, reads "Professor J. Watson has been awarded by the Paris Academy of Sciences, the astronomical prize, Lalande foundation, for the discovery of eight new asteroids in one year. The planet Lydia (No. 110), discovered by M. Borelly at the Marseilles Observatory [...] M. Borelly had previously discovered two planets bearing the numbers 91 and 99 in the system of asteroids revolving between Mars and Jupiter".

243 Ida and its moon Dactyl. Dactyl is the first satellite of an asteroid to be discovered.

Historical methods

Asteroid discovery methods have dramatically improved over the past two centuries.

In the last years of the 18th century, Baron Franz Xaver von Zach organized a group of 24 astronomers to search the sky for the missing planet predicted at about 2.8 AU from the Sun by the Titius-Bode law, partly because of the discovery, by Sir William Herschel in 1781, of the planet Uranus at the distance predicted by the law. This task required that hand-drawn sky charts be prepared for all stars in the zodiacal band down to an agreed-upon limit of faintness. On subsequent nights, the sky would be charted again and any moving object would, hopefully, be spotted. The expected motion of the missing planet was about 30 seconds of arc per hour, readily discernible by observers.

The first object, Ceres, was not discovered by a member of the group, but rather by accident in 1801 by Giuseppe Piazzi, director of the observatory of Palermo in Sicily. He discovered a new star-like object in Taurus and followed the displacement of this object during several nights. His colleague, Carl Friedrich Gauss, used these observations to find the exact distance from this unknown object to the Earth. Gauss' calculations placed the object between the planets Mars and Jupiter. Piazzi named it after Ceres, the Roman goddess of agriculture.

Three other asteroids (2 Pallas, 3 Juno, and 4 Vesta) were discovered over the next few years, with Vesta found in 1807. After eight more years of fruitless searches, most astronomers assumed that there were no more and abandoned any further searches.

However, Karl Ludwig Hencke persisted, and began searching for more asteroids in 1830. Fifteen years later, he found 5 Astraea, the first new asteroid in 38 years. He also found 6 Hebe less than two years later. After this, other astronomers joined in the search and at least one new asteroid was discovered every year after that (except the wartime year 1945). Notable asteroid hunters of this early era were J. R. Hind, Annibale de Gasparis, Robert Luther, H. M. S. Goldschmidt, Jean Chacornac, James Ferguson, Norman Robert Pogson, E. W. Tempel, J. C. Watson, C. H. F. Peters, A. Borrelly, J. Palisa, the Henry brothers and Auguste Charlois.

In 1891, however, Max Wolf pioneered the use of astrophotography to detect asteroids, which appeared as short streaks on long-exposure photographic plates. This dramatically increased the rate of detection compared with earlier visual methods: Wolf alone discovered 248 asteroids, beginning with 323 Brucia, whereas only slightly more than 300 had been discovered up to that point. It was known that there were many more, but most astronomers did not bother with them, calling them "vermin of the skies", a phrase due to Edmund Weiss.[13] Even a century later, only a few thousand asteroids were identified, numbered and named.

Manual methods of the 1900s and modern reporting

Until 1998, asteroids were discovered by a four-step process. First, a region of the sky was photographed by a wide-field telescope, or Astrograph. Pairs of photographs were taken, typically one hour apart. Multiple pairs could be taken over a series of days. Second, the two films or plates of the same region were viewed under a stereoscope. Any body in orbit around the Sun would move slightly between the pair of films. Under the stereoscope, the image of the body would seem to float slightly above the background of stars. Third, once a moving body was identified, its location would be measured precisely using a digitizing microscope. The location would be measured relative to known star locations.[14]

These first three steps do not constitute asteroid discovery: the observer has only found an apparition, which gets a provisional designation, made up of the year of discovery, a letter representing the half-month of discovery, and finally a letter and a number indicating the discovery's sequential number (example: 1998 FJ_{74}).

The last step of discovery is to send the locations and time of observations to the Minor Planet Center, where computer programs determine whether an apparition ties together earlier apparitions into a single orbit. If so, the object receives a catalogue number and the observer of the first apparition with a calculated orbit is declared the discoverer, and granted the honor of naming the object subject to the approval of the International Astronomical Union.

Computerized methods

There is increasing interest in identifying asteroids whose orbits cross Earth's, and that could, given enough time, collide with Earth (see Earth-crosser asteroids). The three most important groups of near-Earth asteroids are the Apollos, Amors, and Atens. Various asteroid deflection strategies have been proposed, as early as the 1960s.

2004 FH is the center dot being followed by the sequence; the object that flashes by during the clip is an artificial satellite.

The near-Earth asteroid 433 Eros had been discovered as long ago as 1898, and the 1930s brought a flurry of similar objects. In order of discovery, these were: 1221 Amor, 1862 Apollo, 2101 Adonis, and finally 69230 Hermes, which approached within 0.005 AU of the Earth in 1937. Astronomers began to realize the possibilities of Earth impact.

Two events in later decades increased the alarm: the increasing acceptance of Walter Alvarez' hypothesis that an impact event resulted in the Cretaceous–Paleogene extinction, and the 1994 observation of Comet Shoemaker-Levy 9 crashing into Jupiter. The U.S. military also declassified the information that its military satellites, built to detect nuclear explosions, had detected hundreds of upper-atmosphere impacts by objects ranging from one to 10 metres across.

All these considerations helped spur the launch of highly efficient automated systems that consist of Charge-Coupled Device (CCD) cameras and computers directly connected to telescopes. Since 1998, a large majority of the asteroids have been discovered by such automated systems. A list of teams using such automated systems includes:[15]

- The Lincoln Near-Earth Asteroid Research (LINEAR) team
- The Near-Earth Asteroid Tracking (NEAT) team
- Spacewatch
- The Lowell Observatory Near-Earth-Object Search (LONEOS) team
- The Catalina Sky Survey (CSS)
- The Campo Imperatore Near-Earth Objects Survey (CINEOS) team
- The Japanese Spaceguard Association
- The Asiago-DLR Asteroid Survey (ADAS)

The LINEAR system alone has discovered 121,346 asteroids, as of March, 2011.[16] Among all the automated systems, 4711 near-Earth asteroids have been discovered[17] including over 600 more than 1 km (0.6 mi) in diameter.

Terminology

Traditionally, small bodies orbiting the Sun were classified as asteroids, comets or meteoroids, with anything smaller than ten metres across being called a meteoroid.[18] The term "asteroid" is ill-defined. It never had a formal definition, with the broader term minor planet being preferred by the International Astronomical Union from 1853 on. In 2006, the term "small Solar System body" was introduced to cover both most minor planets and comets.[19] Other languages prefer "planetoid" (Greek for "planet-like"), and this term is occasionally used in English for the larger asteroids. The word "planetesimal" has a similar meaning, but refers specifically to the small building blocks of the planets that existed when the Solar System was forming. The term "planetule" was coined by the geologist William Daniel Conybeare to describe minor planets,[20] but is not in common use. The three largest objects in the asteroid belt, Ceres, 2 Pallas, and 4 Vesta, grew to the stage of protoplanets. Ceres has been classified as a dwarf planet, the only one in the inner Solar System.

When found, asteroids were seen as a class of objects distinct from comets, and there was no unified term for the two until "small Solar System body" was coined in 2006. The main difference between an asteroid and a comet is that a comet shows a coma due to sublimation of near surface ices by solar radiation. A few objects have ended up being dual-listed because they were first classified as minor planets but later showed evidence of cometary activity. Conversely, some (perhaps all) comets are eventually depleted of their surface volatile ices and become asteroids. A further distinction is that comets typically have more eccentric orbits than most asteroids; most "asteroids" with notably eccentric orbits are probably dormant or extinct comets.[21]

For almost two centuries, from the discovery of Ceres in 1801 until the discovery of the first centaur, 2060 Chiron, in 1977, all known asteroids spent most of their time at or within the orbit of Jupiter, though a few such as 944 Hidalgo ventured far beyond Jupiter for part of their orbit. When astronomers started finding more small bodies that permanently resided further out than Jupiter, now called centaurs, they numbered them among the traditional asteroids, though there was debate over whether they should be classified as asteroids or as a new type of object. Then, when the first trans-Neptunian object, 1992 QB1, was discovered in 1992, and especially when large numbers of similar objects started turning up, new terms were invented to sidestep the issue: Kuiper-belt object, trans-Neptunian object, scattered-disc object, and so on. These inhabit the cold outer reaches of the Solar System where ices remain solid and comet-like bodies are not expected to exhibit much cometary activity; if centaurs or trans-Neptunian objects were to venture close to the Sun, their volatile ices would sublimate, and traditional approaches would classify them as comets and not asteroids.

The innermost of these are the Kuiper-belt objects, called "objects" partly to avoid the need to classify them as asteroids or comets.[22] They are believed to be predominantly comet-like in composition, though some may be more akin to asteroids.[23] Furthermore, most do not have the highly eccentric orbits associated with comets, and the ones so far discovered are larger than traditional comet nuclei. (The much more distant Oort cloud is hypothesized to be the main reservoir of dormant comets.) Other recent observations, such as the analysis of the cometary dust collected by the Stardust probe, are increasingly blurring the distinction between comets and asteroids,[24] suggesting "a continuum between asteroids and comets" rather than a sharp dividing line.[25]

The minor planets beyond Jupiter's orbit are sometimes also called "asteroids", especially in popular presentations.[26] However, it is becoming increasingly common for the term "asteroid" to be restricted to minor planets of the inner Solar System.[22] Therefore, this article will restrict itself for the most part to the classical asteroids: objects of the asteroid belt, Jupiter trojans, and near-Earth objects.

When the IAU introduced the class small Solar System bodies in 2006 to include most objects previously classified as minor planets and comets, they created the class of dwarf planets for the largest minor planets—those that have enough mass to have become ellipsoidal under their own gravity. According to the IAU, "the term 'minor planet' may still be used, but generally the term 'Small Solar System Body' will be preferred."[27] Currently only the largest object in the asteroid belt, Ceres, at about 950 km (590 mi) across, has been placed in the dwarf planet category, although there are several large asteroids (Vesta, Pallas, and Hygiea) that may be classified as dwarf planets when their shapes are better known.[28]

Formation

It is believed that planetesimals in the asteroid belt evolved much like the rest of the solar nebula until Jupiter neared its current mass, at which point excitation from orbital resonances with Jupiter ejected over 99% of planetesimals in the belt. Simulations and a discontinuity in spin rate and spectral properties suggest that asteroids larger than approximately 120 km (75 mi) in diameter accreted during that early era, whereas smaller bodies are fragments from collisions between asteroids during or after the Jovian disruption.[29] Ceres and Vesta grew large enough to melt and differentiate, with heavy metallic elements sinking to the core, leaving rocky minerals in the crust.[30]

In the Nice model, many Kuiper-belt objects are captured in the outer asteroid belt, at distances greater than 2.6 AU. Most were later ejected by Jupiter, but those that remained may be the D-type asteroids, and possibly include Ceres.[31]

Distribution within the Solar System

Various dynamical groups of asteroids have been discovered orbiting in the inner Solar System. Their orbits are perturbed by the gravity of other bodies in the Solar System and by the Yarkovsky effect. Significant populations include:

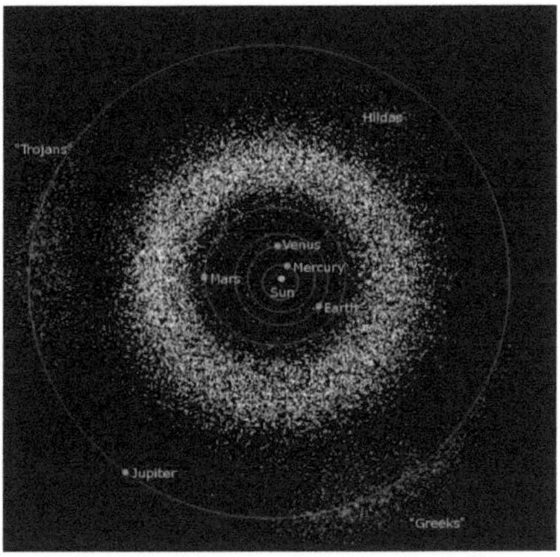

The asteroid belt (white) and the Trojan asteroids (green)

Asteroid belt

The majority of known asteroids orbit within the asteroid belt between the orbits of Mars and Jupiter, generally in relatively low-eccentricity (i.e., not very elongated) orbits. This belt is now estimated to contain between 1.1 and 1.9 million asteroids larger than 1 km (0.6 mi) in diameter,[32] and millions of smaller ones.[33] These asteroids may be remnants of the protoplanetary disk, and in this region the accretion of planetesimals into planets during the formative period of the Solar System was prevented by large gravitational perturbations by Jupiter.

Trojans

Trojan asteroids are a population that share an orbit with a larger planet or moon, but do not collide with it because they orbit in one of the two Lagrangian points of stability, L4 and L5, which lie 60° ahead of and behind the larger body.

The most significant population of Trojan asteroids are the Jupiter Trojans. Although fewer Jupiter Trojans have been discovered as of 2010, it is thought that they are as numerous as the asteroids in the asteroid belt.

A couple of trojans have also been found orbiting with Mars.[34]

Near-Earth asteroids

Near-Earth asteroids, or NEAs, are asteroids that have orbits that pass close to that of Earth. Asteroids that actually cross the Earth's orbital path are known as *Earth-crossers*. As of May 2010, 7,075 near-Earth asteroids are known and the number over one kilometre in diameter is estimated to be 500–1,000.

Characteristics

Size distribution

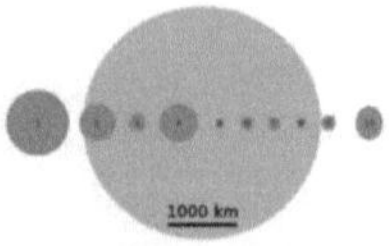

Sizes of the first ten asteroids to be discovered, compared to the Earth's Moon

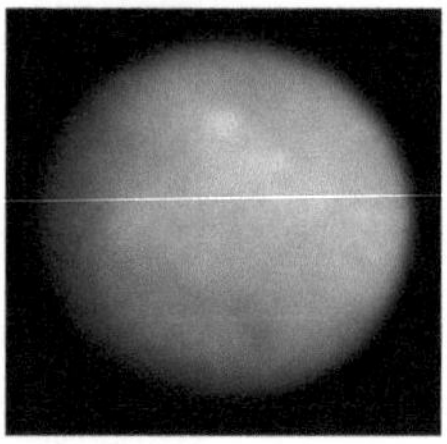

HST image of the dwarf planet Ceres

Asteroids vary greatly in size, from almost 1000 kilometres for the largest down to rocks just tens of metres across.[35] The three largest are very much like miniature planets: they are roughly spherical, have at least partly differentiated interiors,[36] and are thought to be surviving protoplanets. The vast majority, however, are much smaller and are irregularly shaped; they are thought to be either surviving planetesimals or fragments of larger bodies.

The dwarf planet Ceres is by far the largest asteroid, with a diameter of 975 km (610 mi). The next largest are 2 Pallas and 4 Vesta, both with diameters of just over 500 km (300 mi). Vesta is the only main-belt asteroid that can, on occasion, be visible to the naked eye. On some rare occasions, a near-Earth asteroid may briefly become visible without technical aid; see 99942 Apophis.

The mass of all the objects of the asteroid belt, lying between the orbits of Mars and Jupiter, is estimated to be about $2.8–3.2 \times 10^{21}$ kg, or about 4 percent of the mass of the Moon. Of this, Ceres comprises 0.95×10^{21} kg, a third of the total.[37] Adding in the next three most massive objects, Vesta (9%), Pallas (7%), and Hygiea (3%), brings this figure up to 51%; while the three after that, 511 Davida (1.2%), 704 Interamnia (1.0%), and 52 Europa (0.9%), only add another 3% to the total mass. The number of asteroids then increases rapidly as their individual masses decrease.

The number of asteroids decreases markedly with size. Although this generally follows a power law, there are 'bumps' at 5 km and 100 km, where more asteroids than expected from a logarithmic distribution are found.[38]

Approximate number of asteroids N larger than diameter D

D	100 m	300 m	500 m	1 km	3 km	5 km	10 km	30 km	50 km	100 km	200 km	300 km	500 km	900 km
N	~25,000,000	4,000,000	2,000,000	750,000	200,000	90,000	10,000	1,100	600	200	30	5	3	1

Largest asteroids

Although their location in the asteroid belt excludes them from planet status, the four largest objects, Ceres, Vesta, Pallas, and Hygiea, are remnant protoplanets that share many characteristics common to planets, and are atypical compared to the majority of "potato"-shaped asteroids.

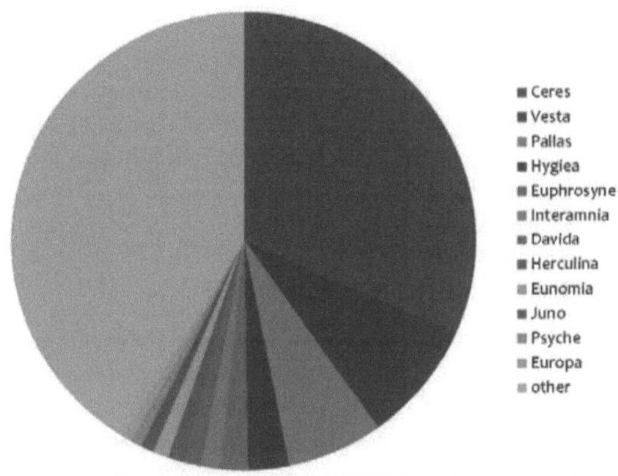

The relative masses of the twelve List of notable asteroids#Largest by masslargest asteroids known, "Recent Asteroid Mass Determinations". Maintained by Jim Baer. Last updated 2010-12-12. Access date 2011-09-02. The values of Juno and Herculina may be off by as much as 16%, and Euphrosyne by a third. The order of the lower eight may change as better data is acquired, but the values do not overlap with any known asteroid outside these twelve. compared to the remaining mass of the asteroid belt. Elena V. PitjevaPitjeva, E. V. (2005). "High-Precision Ephemerides of Planets—EPM and Determination of Some Astronomical Constants" (PDF). Solar System Research 39 (3): 184. Bibcode 2005SoSyR..39..176P. doi:10.1007/s11208-005-0033-2. .

Attributes of protoplanetary asteroids

Name	Orbital radius (AU)	Orbital period (years)	Inclination to ecliptic	Orbital eccentricity	Diameter (km)	Diameter (% of Moon)	Mass ($\times10^{18}$ kg)	Mass (% of Ceres)	Density[41] g/cm^3	Rotation period (hr)	Axial tilt	Surface temperature
Vesta	2.36	3.63	7.1°	0.089	573×557×446 (mean 525)	15%	260	28%	3.44 ± 0.12	5.34	29°	85–270 K
Ceres	2.77	4.60	10.6°	0.079	975×975×909 (mean 952)	28%	940	100%	2.12 ± 0.04	9.07	≈ 3°	167 K
Pallas	2.77	4.62	34.8°	0.231	580×555×500 (mean 545)	16%	210	22%	2.71 ± 0.11	7.81	≈ 80°	164 K
Hygiea	3.14	5.56	3.8°	0.117	530×407×370 (mean 430)	12%	87	9%	2.76 ± 1.2	27.6	≈ 60°	164 K

Ceres is the only asteroid large enough for its gravity to force it into a spheroidal shape, and so, according to the IAU's 2006 resolution on the definition of a planet, it has been classified as a dwarf planet.[42] Vesta may eventually be so classified as well. Ceres has a much higher absolute magnitude than the other asteroids, of around 3.32,[43] and may possess a surface layer of ice.[44] Like the planets, Ceres is differentiated: it has a crust, a mantle and a core.[44] Vesta, too, has a differentiated interior, though it formed inside the Solar System's frost line, and so is devoid of water;[45] its composition is mainly of basaltic rock such as olivine.[46] Pallas is unusual in that, like Uranus, it rotates on its side, with one pole regularly facing the Sun and the other facing away.[47] Its composition is similar to that of Ceres: high in carbon and silicon, and perhaps partially differentiated.[48] Hygiea is a carbonaceous asteroid and, unlike the other largest asteroids, lies relatively close to the plane of the ecliptic.[49]

Rotation

Measurements of the rotation rates of large asteroids in the asteroid belt show that there is an upper limit. No asteroid with a diameter larger than 100 meters has a rotation period smaller than 2.2 hours. For asteroids rotating faster than approximately this rate, the inertia at the surface is greater than the gravitational force, so any loose surface material would be flung out. However, a solid object should be able to rotate much more rapidly. This suggests that most asteroids with a diameter over 100 meters are rubble piles formed through accumulation of debris after collisions between asteroids.[50]

Composition

The physical composition of asteroids is varied and in most cases poorly understood. Ceres appears to be composed of a rocky core covered by an icy mantle, where Vesta is thought to have a nickel-iron core, olivine mantle, and basaltic crust.[51] 10 Hygiea, however, which appears to have a uniformly primitive composition of carbonaceous chondrite, is thought to be the largest undifferentiated asteroid. Most of the smaller asteroids are thought to be piles of rubble held together loosely by gravity, though the largest are probably solid. Some asteroids have moons or are co-orbiting binaries: Rubble piles, moons, binaries, and scattered asteroid families are believed to be the results of collisions that disrupted a parent asteroid.

Asteroids contain traces of amino acids and other organic compounds, and some speculate that asteroid impacts may have seeded the early Earth with the chemicals necessary to initiate life, or may have even brought life itself to Earth. (See also panspermia.)[52] In August 2011, a report, based on NASA studies with meteorites found on Earth, was published suggesting DNA and RNA components (adenine, guanine and related organic molecules) may have been formed on asteroids and comets in outer space.[53] [54] [55]

Only one asteroid, 4 Vesta, which has a reflective surface, is normally visible to the naked eye, and this only in very dark skies when it is favorably positioned. Rarely, small asteroids passing close to Earth may be naked-eye visible for a short time.[56]

Composition is calculated from three primary sources: albedo, surface spectrum, and density. The last can only be determined accurately by observing the orbits of moons the asteroid might have. So far, every asteroid with moons has turned out to be a rubble pile, a loose conglomeration of rock and metal that may be half empty space by volume. The investigated asteroids are as large as 280 km in diameter, and include 121 Hermione (268×186×183 km), and 87 Sylvia (384×262×232 km). Only half a dozen asteroids are larger than 87 Sylvia, though none of them have moons; however, some smaller asteroids are thought to be more massive, suggesting they may not have been disrupted, and indeed 511 Davida, the same size as Sylvia to within measurement error, is estimated to be two and a half times as massive, though this is highly uncertain. The fact that such large asteroids as Sylvia can be rubble piles, presumably due to disruptive impacts, has important consequences for the formation of the Solar system: Computer simulations of collisions involving solid bodies show them destroying each other as often as merging, but colliding rubble piles are more likely to merge. This means that the cores of the planets could have formed relatively quickly.[57]

Surface features

253 Mathilde, a C-type asteroid measuring about 50 kilometres (31.07 mi) across, covered in craters half that size. Photograph taken in 1997 by the NEAR Shoemaker probe.

Most asteroids outside the big four (Ceres, Pallas, Vesta, and Hygiea) are likely to be broadly similar in appearance, if irregular in shape. 50-km 253 Mathilde (shown at right) is a rubble pile saturated with craters with diameters the size of the asteroid's radius, and Earth-based observations of 300-km 511 Davida, one of the largest asteroids after the big four, reveal a similarly angular profile, suggesting it is also saturated with radius-size craters.[58] Medium-sized asteroids such as Mathilde and 243 Ida that have been observed up close also reveal a deep regolith covering the surface. Of the big four, Pallas and Hygiea are practically unknown. Vesta has compression fractures encircling a radius-size crater at its south pole but is otherwise a spheroid. Ceres seems quite different in the glimpses Hubble has provided, with surface features that are unlikely to be due to simple craters and impact basins, but details will not be known until *Dawn* arrives in 2015.

Classification

Asteroids are commonly classified according to two criteria: the characteristics of their orbits, and features of their reflectance spectrum.

Orbital classification

Many asteroids have been placed in groups and families based on their orbital characteristics. Apart from the broadest divisions, it is customary to name a group of asteroids after the first member of that group to be discovered. Groups are relatively loose dynamical associations, whereas families are tighter and result from the catastrophic break-up of a large parent asteroid sometime in the past.[59] Families have only been recognized within the asteroid belt. They were first recognised by Kiyotsugu Hirayama in 1918 and are often called Hirayama families in his honor.

About 30% to 35% of the bodies in the asteroid belt belong to dynamical families each thought to have a common origin in a past collision between asteroids. A family has also been associated with the plutoid dwarf planet Haumea.

Quasi-satellites and horseshoe objects

Some asteroids have unusual horseshoe orbits that are co-orbital with the Earth or some other planet. Examples are 3753 Cruithne and 2002 AA_{29}. The first instance of this type of orbital arrangement was discovered between Saturn's moons Epimetheus and Janus.

Sometimes these horseshoe objects temporarily become quasi-satellites for a few decades or a few hundred years, before returning to their earlier status. Both Earth and Venus are known to have quasi-satellites.

Such objects, if associated with Earth or Venus or even hypothetically Mercury, are a special class of Aten asteroids. However, such objects could be associated with outer planets as well.

Spectral classification

This picture of 433 Eros shows the view looking from one end of the asteroid across the gouge on its underside and toward the opposite end. Features as small as 35 m (115 ft) across can be seen.

In 1975, an asteroid taxonomic system based on colour, albedo, and spectral shape was developed by Clark R. Chapman, David Morrison, and Ben Zellner.[60] These properties are thought to correspond to the composition of the asteroid's surface material. The original classification system had three categories: C-types for dark carbonaceous objects (75% of known asteroids), S-types for stony (silicaceous) objects (17% of known asteroids) and U for those that did not fit into either C or S. This classification has since been expanded to include many other asteroid types. The number of types continues to grow as more asteroids are studied.

The two most widely used taxonomies now used are the Tholen classification and SMASS classification. The former was proposed in 1984 by David J. Tholen, and was based on data collected from an eight-color asteroid survey performed in the 1980s. This resulted in 14 asteroid categories.[61] In 2002, the Small Main-Belt Asteroid Spectroscopic Survey resulted in a modified version of the Tholen taxonomy with 24 different types. Both systems have three broad categories of C, S, and X asteroids, where X consists of mostly metallic asteroids, such as the M-type. There are also several smaller classes.[62]

Note that the proportion of known asteroids falling into the various spectral types does not necessarily reflect the proportion of all asteroids that are of that type; some types are easier to detect than others, biasing the totals.

Problems

Originally, spectral designations were based on inferences of an asteroid's composition.[63] However, the correspondence between spectral class and composition is not always very good, and a variety of classifications are in use. This has led to significant confusion. While asteroids of different spectral classifications are likely to be composed of different materials, there are no assurances that asteroids within the same taxonomic class are composed of similar materials.

At present, the spectral classification based on several coarse resolution spectroscopic surveys in the 1990s is still the standard. Scientists cannot agree on a better taxonomic system, largely due to the difficulty of obtaining detailed measurements consistently for a large sample of asteroids (e.g. finer resolution spectra, or non-spectral data such as densities would be very useful).

Exploration

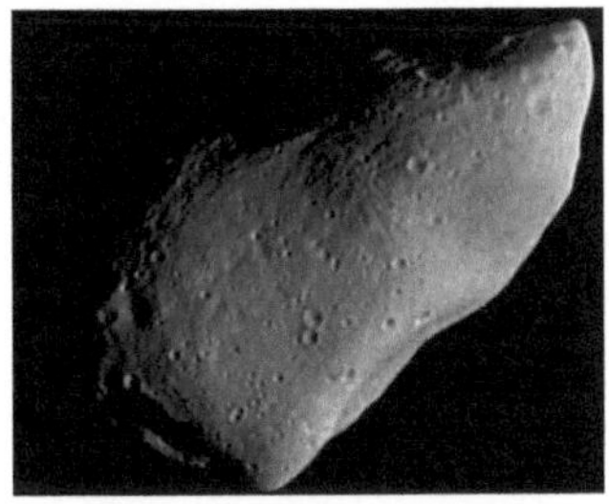
951 Gaspra is the first asteroid to be imaged in close-up.

Until the age of space travel, objects in the asteroid belt were merely pinpricks of light in even the largest telescopes and their shapes and terrain remained a mystery. The best modern ground-based telescopes and the Earth-orbiting Hubble Space Telescope can resolve a small amount of detail on the surfaces of the largest asteroids, but even these mostly remain little more than fuzzy blobs. Limited information about the shapes and compositions of asteroids can be inferred from their light curves (their variation in brightness as they rotate) and their spectral properties, and asteroid sizes can be estimated by timing the lengths of star occulations (when an asteroid passes directly in front of a star). Radar imaging can yield good information about asteroid shapes and orbital and rotational parameters, especially for near-Earth asteroids. In terms of delta-v and propellant requirements, NEOs are more easily accessible than the Moon.[64]

Vesta, imaged by the Dawn spacecraft

The first close-up photographs of asteroid-like objects were taken in 1971 when the Mariner 9 probe imaged Phobos and Deimos, the two small moons of Mars, which are probably captured asteroids. These images revealed the irregular, potato-like shapes of most asteroids, as did later images from the Voyager probes of the small moons of the gas giants.

The first true asteroid to be photographed in close-up was 951 Gaspra in 1991, followed in 1993 by 243 Ida and its moon Dactyl, all of which were imaged by the Galileo probe en route to Jupiter.

The first dedicated asteroid probe was NEAR Shoemaker, which photographed 253 Mathilde in 1997, before entering into orbit around 433 Eros, finally landing on its surface in 2001.

Other asteroids briefly visited by spacecraft en route to other destinations include 9969 Braille (by Deep Space 1 in 1999), and 5535 Annefrank (by Stardust in 2002).

In September 2005, the Japanese Hayabusa probe started studying 25143 Itokawa in detail and was plagued with difficulties, but returned samples of its surface to earth on June 13, 2010.

The European Rosetta probe (launched in 2004) flew by 2867 Šteins in 2008 and 21 Lutetia, the second-largest asteroid visited to date, in 2010.

In September 2007, NASA launched the Dawn Mission, which orbited the protoplanet 4 Vesta from July 2011 to September 2012, and is planned to orbit 1 Ceres in 2015. 4 Vesta is the largest asteroid visited to date.

On 13 December 2012, China's lunar orbiter Chang'e 2 flew within 2 miles (3 km) of the asteroid 4179 Toutatis on an extended mission.

The Japan Aerospace Exploration Agency (JAXA) plans to launch around 2015 the improved Hayabusa 2 space probe and to return asteroid samples by 2020. Current target for the mission is the C-type asteroid (162173) 1999 JU_3.

In May 2011, NASA announced the OSIRIS-REx sample return mission to asteroid 1999 RQ36, and is expected to launch in 2016.

It has been suggested that asteroids might be used as a source of materials that may be rare or exhausted on earth (asteroid mining), or materials for constructing space habitats (see Colonization of the asteroids). Materials that are heavy and expensive to launch from earth may someday be mined from asteroids and used for space manufacturing and construction.

Fiction

Asteroids and the asteroid belt are a staple of science fiction stories. Asteroids play several potential roles in science fiction: as places human beings might colonize, resources for extracting minerals, hazards encountered by spaceships traveling between two other points, and as a threat to life on Earth by potential impact.

See also

- BOOTES (Burst Observer and Optical Transient Exploring System)
- Asteroid deflection strategies
- Meanings of asteroid names
- Pronunciation of asteroid names
- List of asteroids named after people
- List of asteroids named after places
- Category:Asteroid groups and families
- Category:Binary asteroids
- Category:Asteroids
- List of minor planets
- List of notable asteroids
- Centaur (minor planet)
- Dwarf planet
- Impact event
- Lost asteroid
- Mesoplanet
- Minor planet
- Minor Planet Center
- Marco Polo (spacecraft)
- Near-Earth object
- Orion asteroid mission
- Pioneer 10 space probe
- Rosetta probe

Notes

[1] "Asteroids" (http://ssd.jpl.nasa.gov/?asteroids). NASA – Jet Propulsion Laboratory. . Retrieved 13 September2010.

[2] Asimov, Isaac, and Dole, Stephen H. *Planets for Man* (New York: Random House, 1964), p.43.

[3] "What Are Asteroids And Comets?" (http://neo.jpl.nasa.gov/faq/#ast). *Near Earth Object Program FAQ*. NASA. Archived (http://web.archive.org/web/20100909210213/http://neo.jpl.nasa.gov/faq/) from the original on 9 September 2010. . Retrieved 2010-09-13.

[4] Gould, B. A. (1852). "On the Symbolic Notation of the Asteroids". *Astronomical Journal* **2**: 80. Bibcode 1852AJ......2...80G. doi:10.1086/100212.

[5] Hilton, James L. (2001-09-17). "When Did the Asteroids Become Minor Planets" (http://aa.usno.navy.mil/faq/docs/minorplanets.php). . Retrieved 2006-03-26.

[6] Encke, J. F. (1854). "Beobachtung der Bellona, nebst Nachrichten über die Bilker Sternwarte". *Astronomische Nachrichten* **38** (9): 143. doi:10.1002/asna.18540380907.

[7] Rümker, G. (1855). "Name und Zeichen des von Herrn R. Luther zu Bilk am 19. April entdeckten Planeten". *Astronomische Nachrichten* **40** (24): 373. doi:10.1002/asna.18550402405.

[8] Luther, R. (1856). "Schreiben des Herrn Dr. R. Luther, Directors der Sternwarte zu Bilk, an den Herausgeber". *Astronomische Nachrichten* **42** (7): 107. Bibcode 1855AN.....42..107L. doi:10.1002/asna.18550420705.

[9] "When did the asteroids become minor planets?" (http://www.usno.navy.mil/USNO/astronomical-applications/astronomical-information-center/minor-planets). Naval Meteorology and Oceanography Command. . Retrieved 2011-11-06.

[10] Except for Pluto and, in the astrological community, for a few outer bodies such as 2060 Chiron

[11] Ceres is the largest asteroid and is now classified as a dwarf planet. All other asteroids are now classified as small Solar System bodies along with comets, centaurs, and the smaller trans-Neptunian objects.

[12] http://books.google.ca/books?id=NAMAAAAAMAAJ&pg=PA316&dq=%22planets%22+asteroids

[13] Seares, Frederick H. (1930). "Address of the Retiring President of the Society in Awarding the Bruce Medal to Professor Max Wolf". *Publ. Astr. Soc. Pacific* **42**: 5–22. Bibcode 1930PASP...42....5S. doi:10.1086/123986+(Dead+links).

[14] Chapman, Mary G. (May 17, 1992). "Carolyn Shoemaker, Planetary Astronomer and Most Successful 'Comet Hunter' To Date" (http://astrogeology.usgs.gov/About/People/CarolynShoemaker). USGS. . Retrieved 2008-04-15.

[15] Yeomans, Don. "Near Earth Object Search Programs" (http://neo.jpl.nasa.gov/programs/). NASA. Archived (http://web.archive.org/web/20080424093951/http://neo.jpl.nasa.gov/programs/) from the original on 24 April 2008. . Retrieved 2008-04-15.

[16] "Minor Planet Discover Sites" (http://www.minorplanetcenter.org/iau/lists/MPDiscSites.html). Archived (http://web.archive.org/web/20100830074958/http://www.minorplanetcenter.org/iau/lists/MPDiscSites.html) from the original on 30 August 2010. . Retrieved 2010-08-24.

[17] "Unusual Minor Planets" (http://www.minorplanetcenter.org/iau/lists/Unusual.html). Archived (http://web.archive.org/web/20100827234833/http://www.minorplanetcenter.org/iau/lists/Unusual.html) from the original on 27 August 2010. . Retrieved 2010-08-24.

[18] Beech, M. (September 1995). "On the Definition of the Term Meteoroid". *Quarterly Journal of the Royal Astronomical Society* **36** (3): 281–284. Bibcode 1995QJRAS..36..281B.

[19] The definition of "small Solar System bodies" says that they "include most of the Solar System asteroids, most trans-Neptunian objects, comets, and other small bodies". *The Final IAU Resolution on the definition of "planet" ready for voting* (http://www.iau2006.org/mirror/www.iau.org/iau0602/index.html) (IAU)

[20] "English Dictionary – Browsing Page P-44" (http://www.hyperdictionary.com/dict-e/p-44.html). HyperDictionary.com. . Retrieved 2008-04-15.

[21] Weissman, Paul R., William F. Bottke, Jr., and Harold F. Levinson. "Evolution of Comets into Asteroids." *Southwest Research Institute, Planetary Science Directorate.* 2002. Web Retrieved 3 Aug. 2010 (http://www.boulder.swri.edu/~hal/PDF/asteroids3.pdf)

[22] "Are Kuiper Belt Objects asteroids?" (http://curious.astro.cornell.edu/question.php?number=601), "Ask an astronomer", Cornell University

[23] "Asteroids and Comets" (http://rst.gsfc.nasa.gov/Sect19/Sect19_22.html), NASA website

[24] "Comet Dust Seems More Asteroidy" (http://www.sciam.com/podcast/episode.cfm?id=ADD0878B-D6C3-3B70-7B5BC373545BB82D) *Scientific American*, January 25, 2008

[25] "Comet samples are surprisingly asteroid-like" (http://space.newscientist.com/channel/solar-system/comets-asteroids/dn13224-comet-samples-are-surprisingly-asteroidlike.html), *New Scientist*, 24 January 2008

[26] For instance, a joint NASA-JPL public-outreach website states:

> "We include Trojans (bodies captured in Jupiter's 4th and 5th Lagrange points), Centaurs (bodies in orbit between Jupiter and Neptune), and trans-Neptunian objects (orbiting beyond Neptune) in our definition of "asteroid" as used on this site, even though they may more correctly be called "minor planets" instead of asteroids."

<http://ssd.jpl.nasa.gov/?asteroids>

[27] Questions and Answers on Planets (http://www.iau.org/public_press/news/release/iau0603/questions_answers/), IAU

[28] "Three new planets may join solar system" (http://space.newscientist.com/article.ns?id=dn9761&feedId=online-news_rss20F53), *New Scientist*, 16 August 2006

[29] Bottke, Durda; Nesvorny, Jedicke; Morbidelli, Vokrouhlicky; Levison (2005). "The fossilized size distribution of the main asteroid belt" (http://astro.mff.cuni.cz/davok/papers/fossil05.pdf). *Icarus* **175**: 111. Bibcode 2005Icar..175..111B. doi:10.1016/j.icarus.2004.10.026. .

[30] Kerrod, Robin (2000). *Asteroids, Comets, and Meteors*. Lerner Publications Co.. ISBN 0-585-31763-1.

[31] William B. McKinnon, 2008, "On The Possibility Of Large KBOs Being Injected Into The Outer Asteroid Belt". (http://adsabs.harvard.edu/abs/2008DPS....40.3803M) *American Astronomical Society*, DPS meeting #40, #38.03

[32] Tedesco, Edward; Metcalfe, Leo (April 4, 2002). "New study reveals twice as many asteroids as previously believed" (http://www.spaceref.com/news/viewpr.html?pid=7925) (Press release). European Space Agency. . Retrieved 2008-02-21.

[33] World Book at NASA (http://www.nasa.gov/worldbook/asteroid_worldbook.html)

[34] Neptune also has a few known trojans, and these are thought to actually be much more numerous than the Jovian trojans. However, they are often included in the trans-Neptunian population rather than counted with the asteroids.

[35] Below 10 metres, these rocks are by convention considered to be meteoroids.

[36] Schmidt, B.; Russell, C. T.; Bauer, J. M.; Li, J.; McFadden, L. A.; Mutchler, M.; Parker, J. W.; Rivkin, A. S.; Stern, S. A.; Thomas, P. C. (2007). "Hubble Space Telescope Observations of 2 Pallas". *American Astronomical Society, DPS meeting #39* **39**: 485. Bibcode 2007DPS....39.3519S.

[37] Pitjeva, E. V. (2004). "Estimations of masses of the largest asteroids and the main asteroid belt from ranging to planets, Mars orbiters and landers" (http://adsabs.harvard.edu/abs/2004cosp.meet.2014P). *35th COSPAR Scientific Assembly. Held 18–25 July 2004, in Paris, France*. pp. 2014. .

[38] Davis 2002, "Asteroids III", cited by Željko Ivezić (http://www.astro.washington.edu/users/ivezic/Astr598/lecture4.pdf)

[39] "Recent Asteroid Mass Determinations" (http://home.earthlink.net/~jimbaer1/astmass.txt). Maintained by Jim Baer. Last updated 2010-12-12. Access date 2011-09-02. The values of Juno and Herculina may be off by as much as 16%, and Euphrosyne by a third. The order of the lower eight may change as better data is acquired, but the values do not overlap with any known asteroid outside these twelve.

[40] Pitjeva, E. V. (2005). "High-Precision Ephemerides of Planets—EPM and Determination of Some Astronomical Constants" (http://iau-comm4.jpl.nasa.gov/EPM2004.pdf) (PDF). *Solar System Research* **39** (3): 184. Bibcode 2005SoSyR..39..176P. doi:10.1007/s11208-005-0033-2. .

[41] "Asteroid Density, Porosity, and Structure" (http://www.lpi.usra.edu/books/AsteroidsIII/pdf/3022.pdf). lpi.usra.edu. . Retrieved 2013-01-03.

[42] "The Final IAU Resolution on the Definition of "Planet" Ready for Voting" (http://www.iau.org/public_press/news/detail/iau0602/). IAU. August 24, 2006. . Retrieved 2007-03-02.

[43] Parker, J. W.; Stern, S. A.; Thomas, P. C.; Festou, M. C.; Merline, W. J.; Young, E. F.; Binzel, R. P.; and Lebofsky, L. A. (2002). "Analysis of the First Disk-resolved Images of Ceres from Ultraviolet Observations with the Hubble Space Telescope". *The Astronomical Journal* **123** (1): 549–557. arXiv:astro-ph/0110258. Bibcode 2002AJ....123..549P. doi:10.1086/338093.

[44] "Asteroid 1 Ceres" (http://www.planetary.org/explore/topics/asteroids_and_comets/ceres.html). *The Planetary Society*. Archived (http://web.archive.org/web/20070929092440/http://planetary.org/explore/topics/asteroids_and_comets/ceres.html) from the original

on 29 September 2007. . Retrieved 2007-10-20.

[45] "Key Stages in the Evolution of the Asteroid Vesta" (http://hubblesite.org/newscenter/newsdesk/archive/releases/1995/20/image/c). *Hubble Space Telescope news release*. 1995. Archived (http://web.archive.org/web/20070930203613/http://hubblesite.org/newscenter/newsdesk/archive/releases/1995/20/image/c) from the original on 30 September 2007. . Retrieved 2007-10-20. Russel, C. T.; *et al.* (2007). "Dawn mission and operations" (http://journals.cambridge.org/action/displayAbstract?fromPage=online&aid=414750). *NASA/JPL*. . Retrieved 2007-10-20.

[46] Burbine, T. H. (July 1994). "Where are the olivine asteroids in the main belt?" (http://articles.adsabs.harvard.edu/cgi-bin/nph-iarticle_query?1994Metic..29..453B&data_type=PDF_HIGH&whole_paper=YES&type=PRINTER&filetype=.pdf). *Meteoritics* **29** (4): 453. Bibcode 1994Metic..29..453B. .

[47] Torppa, J.; *et al.* (1996). "Shapes and rotational properties of thirty asteroids from photometric data". *Icarus* **164** (2): 346–383. Bibcode 2003Icar..164..346T. doi:10.1016/S0019-1035(03)00146-5.

[48] Larson, H. P.; Feierberg, M. A.; and Lebofsky, L. A. (1983). "The composition of asteroid 2 Pallas and its relation to primitive meteorites" (http://adsabs.harvard.edu/abs/1983Icar...56..398L). . Retrieved 2007-10-20.

[49] Barucci, M. A.; *et al.* (2002). "10 Hygiea: ISO Infrared Observations" (http://www.lesia.obspm.fr/~crovisier/biblio/preprint/bar02_icarus.pdf) (PDF). Archived (http://web.archive.org/web/20071128200223/http://www.lesia.obspm.fr/~crovisier/biblio/preprint/bar02_icarus.pdf) from the original on 28 November 2007. . Retrieved 2007-10-21. "Ceres the Planet" (http://www.orbitsimulator.com/gravity/articles/ceres.html). *orbitsimulator.com*. Archived (http://web.archive.org/web/20071011154140/http://orbitsimulator.com/gravity/articles/ceres.html) from the original on 11 October 2007. . Retrieved 2007-10-20.

[50] Rossi, Alessandro (2004-05-20). "The mysteries of the asteroid rotation day" (http://spaceguard.iasf-roma.inaf.it/tumblingstone/issues/current/eng/ast-day.htm). The Spaceguard Foundation. . Retrieved 2007-04-09.

[51] HubbleSite - NewsCenter - Asteroid or Mini-Planet? Hubble Maps the Ancient Surface of Vesta (04/19/1995) - Release Images (http://hubblesite.org/newscenter/archive/releases/1995/20/image/)

[52] Life is Sweet: Sugar-Packing Asteroids May Have Seeded Life on Earth (http://www.space.com/scienceastronomy/planetearth/meteor_sugar_011219.html) Archived (http://web.archive.org/20020124092631/http://web.archive.org/web/20020124092631/http://www.space.com/scienceastronomy/planetearth/meteor_sugar_011219.html) January 24, 2002 at the Wayback Machine, Space.com, 19 December 2001

[53] Callahan, M.P.; Smith, K.E.; Cleaves, H.J.; Ruzica, J.; Stern, J.C.; Glavin, D.P.; House, C.H.; Dworkin, J.P. (11 August 2011). "Carbonaceous meteorites contain a wide range of extraterrestrial nucleobases" (http://www.pnas.org/content/early/2011/08/10/1106493108). PNAS. doi:10.1073/pnas.1106493108. . Retrieved 2011-08-15.

[54] Steigerwald, John (8 August 2011). "NASA Researchers: DNA Building Blocks Can Be Made in Space" (http://www.nasa.gov/topics/solarsystem/features/dna-meteorites.html). NASA. . Retrieved 2011-08-10.

[55] ScienceDaily Staff (9 August 2011). "DNA Building Blocks Can Be Made in Space, NASA Evidence Suggests" (http://www.sciencedaily.com/releases/2011/08/110808220659.htm). ScienceDaily. . Retrieved 2011-08-09.

[56] Closest Flyby of Large Asteroid to be Naked-Eye Visible (http://www.space.com/spacewatch/050204_2004_mn4.html), Space.com, 4 February 2005

[57] Marchis, Descamps, et al. *Icarus*, Feb. 2011

[58] A.R. Conrad *et al.* 2007. " Direct measurement of the size, shape, and pole of 511 Davida with Keck AO in a single night (http://www2.keck.hawaii.edu/inst/people/conrad/research/pub/d511.pdf)", *Icarus,* doi:10.1016/j.icarus.2007.05.004

[59] Zappalà, V. (1995). "Asteroid families: Search of a 12,487-asteroid sample using two different clustering techniques". *Icarus* **116** (2): 291–314. Bibcode 1995Icar..116..291Z. doi:10.1006/icar.1995.1127.

[60] Chapman, C. R. (1975). "Surface properties of asteroids: A synthesis of polarimetry, radiometry, and spectrophotometry". *Icarus* **25** (1): 104–130. Bibcode 1975Icar...25..104C. doi:10.1016/0019-1035(75)90191-8.

[61] Tholen, D. J. (March 8–11, 1988). "Asteroid taxonomic classifications" (http://adsabs.harvard.edu/abs/1989aste.conf.1139T). *Asteroids II; Proceedings of the Conference*. Tucson, AZ: University of Arizona Press. pp. 1139–1150. . Retrieved 2008-04-14.

[62] Bus, S. J. (2002). "Phase II of the Small Main-belt Asteroid Spectroscopy Survey: A feature-based taxonomy". *Icarus* **158** (1): 146. Bibcode 2002Icar..158..146B. doi:10.1006/icar.2002.6856.

[63] McSween Jr., Harry Y. (1999). *Meteorites and their Parent Planets* (2nd ed.). Oxford University Press. ISBN 0-521-58751-4.

[64] A Piloted Orion Flight to a Near-Earth Object: A Feasibility Study (http://ti.arc.nasa.gov/m/project/neo/pdf/NEO_feasibility.pdf)

References

External links

- Rocks from the Main Belt asteroids (http://rocksfromspace.open.ac.uk/asteroid_belt_detail.htm)
- Alphabetical list of minor planet names (ASCII) (http://www.minorplanetcenter.org/iau/lists/MPNames.html) (Minor Planet Center)
- Near Earth Asteroid Tracking (NEAT) (http://neat.jpl.nasa.gov/)
- Asteroids Page (http://solarsystem.nasa.gov/planets/profile.cfm?Object=Asteroids) at NASA's Solar System Exploration (http://solarsystem.nasa.gov/)
- Asteroid Simulator with Moon and Earth (http://www.colorado.edu/physics/2000/applets/satellites.html)
- Alphabetical and numerical lists of minor planet names (Unicode) (http://www.ipa.nw.ru/PAGE/DEPFUND/LSBSS/englenam.htm) (Institute of Applied Astronomy)
- Future Asteroid Interception Research (http://www.fair-society.org/)
- Near Earth Objects Dynamic Site (http://newton.dm.unipi.it/cgi-bin/neodys/neoibo)
- Asteroids Dynamic Site (http://hamilton.dm.unipi.it/astdys/) Up-to-date osculating orbital elements and proper orbital elements University of Pisa, Italy.
- JPL small bodies database (http://ssd.jpl.nasa.gov/?sb_elem) Current down-loadable ASCII table of orbit data and absolute mags H for over 200000 asteroids, sorted by number. Caltech/JPL.
- Asteroid naming statistics (http://quasar.ipa.nw.ru/PAGE/DEPFUND/LSBSS/statmpn.htm)
- Spaceguard UK (http://www.spaceguarduk.com/)
- Committee on Small Body Nomenclature (http://www.ss.astro.umd.edu/IAU/csbn/)
- List of minor planet orbital groupings and families from ProjectPluto (http://www.projectpluto.com/mp_group.htm)
- Cunningham, Clifford, "Introduction to Asteroids: The Next Frontier", ISBN 0-943396-16-6
- James L. Hilton: When Did the Asteroids Become Minor Planets? (http://aa.usno.navy.mil/faq/docs/minorplanets.php)
- Kirkwood, Daniel; *Relations between the Motions of some of the Minor Planets* (1874). (http://adsabs.harvard.edu/cgi-bin/nph-bib_query?bibcode=1874MNRAS..35...61K&db_key=AST&high=40daf3f6f901929)
- Schmadel, L.D. (2003). *Dictionary of Minor Planet Names.* 5th ed. IAU/Springer-Verlag: Heidelberg.
- Asteroid articles in Planetary Science Research Discoveries (http://www.psrd.hawaii.edu/Archive/Archive-Asteroids.html)
- Catalogue of the Solar System Small Bodies Orbital Evolution (http://smallbodies.ru/en/)
- TECA Table of next close approaches to the Earth (http://www.brera.mi.astro.it/sormano/teca.html)
- SAEL Small Asteroids Encounter List (http://www.brera.mi.astro.it/sormano/sael.html)
- MBPL Minor Body Priority List (http://www.brera.mi.astro.it/sormano/mbpl.html)
- PCEL Planetary Close Encounter List (http://www.brera.mi.astro.it/sormano/pcel.html)

Edward_L._G._Bowell

Edward L. G. "Ted" Bowell (born 1943 in London), is an American astronomer. Bowell was educated at Emanuel School London, University College, London, and the Université de Paris.

He was principal investigator of the Lowell Observatory Near-Earth-Object Search (LONEOS).

He has discovered a large number of asteroids, both as part of LONEOS and in his own right before LONEOS began. Among the latter are the Trojan asteroids 2357 Phereclos, 2759 Idomeneus, 2797 Teucer, 2920 Automedon, 3564 Talthybius, 4057 Demophon, and (4489) 1988 AK.

He also co-discovered the periodic comet 140P/Bowell-Skiff and the non-periodic comet C/1980 E1.

References

External links

- "Edward Bowell, Astronomer" (http://www.lowell.edu/team/index.php?req=Bowell_Edward). Lowell Observatory.

Anderson_Mesa_Station

Lowell Observatory, Anderson Mesa Station	
Organization	Lowell Observatory, United States Naval Observatory
Code	688
Location	Coconino County, near Flagstaff, Arizona
Coordinates	35°05′49″N 111°32′09″W
Altitude	2163 meters (ft)
Established	1959
Website Lowell Observatory [1]	

Telescopes	
Perkins Telescope	1.8 m Cassegrain telescope
John S. Hall Telescope	1.1 m Ritchey-Chretien telescope
NURO Telescope	0.8 m reflecting telescope
Navy Optical Interferometer	Interferometer

Anderson Mesa Station is an astronomical observatory established in 1959 as a dark-sky observing site for Lowell Observatory. It is located in Coconino County, Arizona (USA), about 12 miles southeast of Lowell's main campus on Mars Hill in Flagstaff, Arizona.

Telescopes

Current telescopes

- The 1.83 m (in) Perkins Telescope is shared with Boston University (BU) and Georgia State University.[2] Built in 1931 by Warner & Swasey Company, it was originally located at the Perkins Observatory of Ohio Wesleyan University (OWU) in Delaware, Ohio.[3] [4] It was moved to Anderson Mesa in 1961, and was purchased by Lowell in 1998. Lowell and BU formed a partnership to operate the telescope that year, and GSU joined later.[5] The original 69-inch mirror, which was figured by J. W. Fecker, Inc., was the largest single piece of glass ever cast in America when it was poured.[3] It was replaced in 1965 with the current mirror made of Duran-50 low-expansion glass.[6] [7]
- The 1.07 m (in) John Hall Telescope was built by AstroMechanics and installed at Anderson Mesa in 1970.[8] It was named after former Lowell Observatory director John S. Hall in 1990.[9] In 2004, the Ritchey-Chrétien telescope was upgraded with a new mirror from Hextek, and with other parts.[10]
- The 0.79 m (in) National Undergraduate Research Observatory (NURO) Telescope was built by AstroMechanics and installed in 1964 at Anderson Mesa by the U.S. Geological Survey (USGS) for Project Apollo.[11] It was purchased by Lowell in 1972, and refurbished in 1990.[12] [9] It is used by the NURO consortium for up to 60% of the time, and by Lowell scientists.
- The Navy Precision Optical Interferometer (NPOI) is collaboration of Lowell Observatory, the U.S. Naval Observatory (USNO) Flagstaff Station (NOFS), and the U.S. Naval Research Laboratory (NRL).[13] Construction on the facility began in 1992, and engineering tests began in 1994. The first images were acquired in 1996.[14]

Former telescopes

- The 0.6 m (in) Lowell Observatory Near-Earth-Object Search (LONEOS) Schmidt camera was used to search for asteroids and other near-earth objects. It was built by J. W. Fecker, Inc. in 1939, given to Perkins Observatory in the 1950s, and purchased by Lowell in 1990.[9] Starting in 1992 it was refurbished, and saw first light in the dome that previously held the Lowell Astrograph in 1997.[15] Use of the telescope ended along with the LONEOS project in 2008.
- The 0.33 m (in) Abbot L. Lowell Astrograph, also known as the Pluto Discovery Telescope and informally as the Pluto Camera, is an astrograph built by Alvan Clark & Sons in 1929. In 1930 it was used by Clyde Tombaugh to discover Pluto.[16] In 1971, it was moved a new building at Anderson Mesa, and returned to Mars Hill in 1992.[17] [18]

See also

- United States Naval Observatory Flagstaff Station
- List of astronomical observatories

References

[1] http://www.lowell.edu/
[2] "72-inch Perkins | Telescopes | Research | Lowell Observatory" (http://www.lowell.edu/research_telescopes_perkins.php). Lowell Observatory. . Retrieved 2012-01-11.
[3] "History of Perkins Observatory" (http://www.perkins-observatory.org/history.html#69 inch mirror). Perkins Observatory. . Retrieved 2012-01-11.
[4] Phillips, Earl W., Jr.. "A Short History of Perkins Observatory" (http://www.setileague.org/articles/perkins.htm). The SETI League, Inc.. . Retrieved 2012-01-22.
[5] Horstman, H. S. (1999). "Lowell Observatory, Flagstaff, Arizona 86001. Report for the period 1 Jul 1997 - 30 Jun 1998". *Bulletin of the Astronomical Society* **31**: 179. Bibcode 1999BAAS...31..179H.
[6] Abrahams, Peter. "The Telescopes of Lowell Observatory" (http://home.europa.com/~telscope/lowell.txt). . Retrieved 2012-01-11.
[7] Hall, S. J. (1964). "Lowell Observatory report". *Astronomical Journal* **69**: 684. Bibcode 1964AJ.....69..684H. doi:10.1086/109345.

[8] Hall, J. S. (1971). "Lowell Observatory, Flagstaff, Arizona. Report 1969-1970". *Bulletin of the Astronomical Society* **3**: 125. Bibcode 1971BAAS....3..125H.

[9] Horstman, H. S.; Millis, R. L. (1991). "Lowell Observatory, Flagstaff, Arizona 86001. Report for the period 1 Jul 1989 - 30 Jun 1990". *Bulletin of the Astronomical Society* **23**: 361. Bibcode 1991BAAS...23..361H.

[10] "42-inch Hall Telescope I Telescopes I Resarch I Lowell Observatory" (http://www.lowell.edu/research_telescopes_hall.php). Lowell Observatory. . Retrieved 2012-01-11.

[11] Schaber, Gerald G. (2005). [pubs.usgs.gov/of/2005/1190/of2005-1190.pdf "The U.S. Geological Survey, Branch of Astrogeology—A Chronology of Activities from Conception through the End of Project Apollo (1960- 1973)"]. United States Geological Survey. pubs.usgs.gov/of/2005/1190/of2005-1190.pdf. Retrieved 2012-01-22.

[12] "31-inch NURO Telescope I Telescopes I Resarch I Lowell Observatory" (http://www.lowell.edu/research_telescopes_nuro.php). Lowell Observatory. . Retrieved 2012-01-11.

[13] "NOI Telescope I Telescopes I Resarch I Lowell Observatory" (http://www.lowell.edu/research_telescopes_noi.php). Lowell Observatory. . Retrieved 2012-01-11.

[14] Armstrong, J. T.; Mozurkewich, D.; Rickard, L. J; Hutter, D. J.; Benson, J. A.; Bowers, P. F.; Elias Ii, N. M.; Hummel, C. A. et al. (1998). "The Navy Prototype Optical Interferometer". *The Astrophysical Journal* **496**: 550. Bibcode 1998ApJ...496..550A. doi:10.1086/305365.

[15] "Observing Site" (http://www2.lowell.edu/users/elgb/observing_site.html). Lowell Observatory. . Retrieved 2012-01-11.

[16] "Our History: Discovering Pluto I About Us I Lowell Observatory" (http://www.lowell.edu/about_history_pluto.php). Lowell Observatory. . Retrieved 2012-01-11.

[17] Hall, J. S. (1972). "Lowell Observatory, Flagstaff, Arizona. Observatory report". *Bulletin of the Astronomical Society* **4**: 105. Bibcode 1972BAAS....4..105H.

[18] Horstman, H. S. (1995). "Lowell Observatory, Flagstaff, Arizona 86001. Report for the period 1 Jul 1993 - 30 Jun 1994". *Bulletin of the Astronomical Society* **27**: 267. Bibcode 1995BAAS...27..267H.

External links

- Lowell Observatory website (http://www.lowell.edu/)
- Anderson Mesa Station Clear sky clock (http://cleardarksky.com/c/AdrMsStAZkey.html?1) Weather forecasts for observing conditions.

Minor_planet

A **minor planet** is an astronomical object in direct orbit around the Sun that is neither a dominant planet nor originally classified as a comet. Minor planets can be dwarf planets, asteroids, trojans, centaurs, Kuiper belt objects, and other trans-Neptunian objects.[1] The first minor planet discovered was Ceres in 1801 (although from the time of its discovery until 1851 it was considered to be a planet). The orbits of more than 570,000 objects have been archived at the Minor Planet Center.[2]

The term "minor planet" has been used since the 19th century to describe these objects.[3] The term **planetoid** has also been used, especially for larger (planetary) objects.[4] Historically, the terms *asteroid*, *minor planet*, and *planetoid* have been more or less synonymous,[4] [5] but the issue has been complicated by the discovery of numerous minor planets beyond the orbit of Jupiter and especially Neptune that are not universally considered asteroids.[5] Minor planets seen outgassing may receive a dual classification as a comet.

Before 2006 the International Astronomical Union had officially used the term *minor planet*. During its 2006 meeting, the Union reclassified minor planets and comets into dwarf planets and small Solar System bodies.[6] Objects are called dwarf planets if their self-gravity is sufficient to achieve hydrostatic equilibrium, that is, an ellipsoidal shape, with all other minor planets and comets called "small Solar System bodies".[6] The IAU states: "the term 'minor planet' may still be used, but generally the term 'small solar system body' will be preferred."[7] However, for purposes of numbering and naming, the traditional distinction between minor planet and comet is still followed.

Populations

Hundreds of thousands of minor planets have been discovered within the Solar System, with the 2009 rate of discovery at over 3,000 per month. Of the more than 535,000 registered minor planets, 251,651 have orbits known well enough to be assigned permanent official numbers.[8] [2] Of these, 16,154 have official names[2] . As of September 2010, the lowest-numbered unnamed minor planet is (3708) 1974 FV_1;[9] but there are also some named minor planets above number 240,000.[10]

There are various broad minor-planet populations:

- Asteroids; traditionally, most have been bodies in the inner Solar System.[5]
  - Main-belt asteroids, those following roughly circular orbits between Mars and Jupiter. These are the original and best-known group of asteroids or minor planets.
  - Near-Earth asteroids, those whose orbits take them inside the orbit of Mars. Further subclassification of these, based on orbital distance, is used:[11]
    - Aten asteroids, those that have semi-major axes of less than one Earth orbit. Those Aten asteroids that have their aphelion within Earth's orbit are known as Apohele asteroids;
    - Amor asteroids are those near-Earth asteroids that approach the orbit of the Earth from beyond, but do not cross it. Amor asteroids are further subdivided into four subgroups, depending on where their semimajor axis falls between Earth's orbit and the asteroid belt;
    - Apollo asteroids are those asteroids with a semimajor axis greater than Earth's, while having a perihelion distance of 1.017 AU or less. Like Aten asteroids, Apollo asteroids are Earth-crossers.
  - Earth trojans, asteroids sharing Earth's orbit and gravitationally locked to it. As of 2011, the only one known is 2010 TK_7.[12]
  - Mars trojans, asteroids sharing Mars's orbit and gravitationally locked to it. As of 2007, eight such asteroids are known.[13]
  - Jupiter trojans, asteroids sharing Jupiter's orbit and gravitationally locked to it. Numerically they are estimated to equal the main-belt asteroids.
- Distant minor planets; an umbrella term for minor planets in the outer Solar System.
  - Centaurs, bodies in the outer Solar System between Jupiter and Neptune. They have unstable orbits due to the gravitational influence of the giant planets, and therefore must have come from elsewhere, probably outside Neptune.[14]
  - Neptune trojans, bodies sharing Neptune's orbit and gravitationally locked to it. Although only a handful are known, there is evidence that Neptune trojans are more numerous than either the asteroids in the asteroid belt or the Jupiter trojans.[15]
  - Trans-Neptunian objects, bodies at or beyond the orbit of Neptune, the outermost planet.
    - The Kuiper belt, objects inside an apparent population drop-off approximately 55 AU from the Sun.
      - Classical Kuiper belt objects, also known as cubewanos, are in primordial, relatively circular orbits that are not in resonance with Neptune.
      - Plutinos, bodies like Pluto that are in a 2:3 resonance with Neptune, and other resonant trans-Neptunian objects. This is a rather large group.
    - Scattered disc, objects with aphelia outside the Kuiper belt. These are thought to have been scattered by Neptune.
    - Detached objects such as Sedna, with both aphelia and perihelia outside the Kuiper belt.
    - The Oort Cloud, a hypothetical population thought to be the source of long-period comets that may extend out to 50,000 AU from the Sun.

Naming

A newly discovered minor planet is given a provisional designation (such as 2002 AT_4) consisting of the year of discovery and an alphanumeric code indicating the half-month of discovery and the sequence within that half-month. Once an asteroid's orbit has been confirmed, it is given a number, and later may also be given a name (e.g. 433 Eros). The formal naming convention uses parentheses around the number, but dropping the parentheses is quite common. Informally, it is common to drop the number altogether, or to drop it after the first mention when a name is repeated in running text.

Minor planets that have been given a number but not a name keep their provisional designation, e.g. (29075) 1950 DA. As modern discovery techniques are finding vast numbers of new asteroids, they are increasingly being left unnamed. The earliest discovered to be left unnamed was for a long time (3360) 1981 VA, now 3360 Syrinx; as of September 2008, this distinction is held by (3708) 1974 FV_1. On rare occasions, a small object's provisional designation may become used as a name in itself: the still unnamed (15760) 1992 QB_1 gave its "name" to a group of Kuiper belt objects which became known as cubewanos.

A few objects are cross-listed as both comets and asteroids, such as 4015 Wilson–Harrington, which is also listed as 107P/Wilson–Harrington.

Numbering

Minor planets are awarded an official number once their orbits are confirmed. With the increasing rapidity of discovery, these are now six-figure numbers. The switch from five figures to six figures arrived with the publication of the Minor Planet Circular (MPC) of October 19, 2005, which saw the highest numbered minor planet jump from 99947 to 118161.

Sources for names

The first few asteroids were named after figures from Greek and Roman mythology, but as such names started to dwindle the names of famous people, literary characters, discoverer's wives, children, and even television characters were used.

The first asteroid to be given a non-mythological name was 20 Massalia, named after the Greek name for the city of Marseilles. The first to be given an entirely non-Classical name was 45 Eugenia, named after Empress Eugénie de Montijo, the wife of Napoleon III. For some time only female (or feminized) names were used; Alexander von Humboldt was the first man to have an asteroid named after him, but his name was feminized to 54 Alexandra. This unspoken tradition lasted until 334 Chicago was named; even then, oddly feminised names show up in the list for years after.

As the number of asteroids began to run into the hundreds, and eventually the thousands, discoverers began to give them increasingly frivolous names. The first hints of this were 482 Petrina and 483 Seppina, named after the discoverer's pet dogs. However, there was little controversy about this until 1971, upon the naming of 2309 Mr. Spock (the name of the discoverer's cat). Although the IAU subsequently banned pet names as sources, eccentric asteroid names are still being proposed and accepted, such as 4321 Zero, 6042 Cheshirecat, 9007 James Bond, 13579 Allodd and 24680 Alleven, and 26858 Misterrogers.

A well-established rule is that, unlike comets, minor planets may not be named after their discoverer(s). One way to circumvent this rule has been for astronomers to exchange the courtesy of naming their discoveries after each other. An exception to this rule is 96747 Crespodasilva, which was named after its discoverer, Lucy d'Escoffier Crespo da Silva, because she died shortly after the discovery, at age 22.[16] [17]

Names were adapted to various languages from the beginning. 1 Ceres, *Ceres* being its Anglo-Latin name, was actually named *Cerere*, the Italian form of the name. German, French, Arabic and Hindi use forms similar to the English, whereas Russian uses a form, *Tserera*, similar to the Italian. In Greek the name was translated to Demeter, the Greek equivalent of the Roman goddess Ceres. In the early years, before it started causing conflicts, asteroids named after Roman figures were generally translated in Greek; other examples are Hera for 3 Juno, Hestia for 4 Vesta, Chloris for 8 Flora, and Pistê for 37 Fides. In Chinese, the names are not given the Chinese forms of the deities they are named after, but rather typically have a syllable or two for the character of the deity or person, followed by 'god(dess)' or 'woman' if just one syllable, plus 'star/planet', so that most asteroid names are written with three Chinese characters. Thus Ceres is 'grain goddess planet',[18] Pallas is 'wisdom goddess planet', etc.

Special naming rules

Minor-planet naming is not always a free-for-all: there are some populations for which rules have developed about the sources of names. For instance, centaurs (orbiting between Saturn and Neptune) are all named after mythological centaurs; Jupiter trojans after heroes from the Trojan War; resonant trans-Neptunian objects after underworld spirits; and non-resonant TNOs after creation deities.

Physical properties of comets and minor planets

Commission 15[19] of the International Astronomical Union is dedicated to the Physical Study of Comets & Minor Planets.

Archival data on the physical properties of comets and minor planets are found in the PDS Asteroid/Dust Archive.[20] This includes standard asteroid physical characteristics such as the properties of binary systems, occultation timings and diameters, masses, densities, rotation periods, surface temperatures, albedoes, spin vectors, taxonomy, and absolute magnitudes and slopes. In addition, European Asteroid Research Node (E.A.R.N.), an association of asteroid research groups, maintains a Data Base of Physical and Dynamical Properties of Near Earth Asteroids.[21]

Most detailed information is available from Category: Asteroids visited by spacecraft and Category: Comets visited by spacecraft.

See also

- List of minor planets
- Solar System
- Groups of minor planets
- Small Solar System body
- Plutoid
- Quasi-satellite

External links

- Minor Planet Center [22]

References

[1] "Unusual Minor Planets" (http://www.minorplanetcenter.net/iau/lists/Unusual.html). Minor Planet Center. . Retrieved 23 December 2011.

[2] "Minor Planet Statistics" (http://www.minorplanetcenter.net/iau/lists/ArchiveStatistics.html). Minor Planet Center. . Retrieved 2011-02-26.

[3] When did the asteroids become minor planets? (http://aa.usno.navy.mil/faq/docs/minorplanets.php), James L. Hilton, Astronomical Information Center, United States Naval Observatory. Accessed May 5, 2008.

[4] Planet, asteroid, minor planet: A case study in astronomical nomenclature, David W. Hughes, Brian G. Marsden, *Journal of Astronomical History and Heritage* **10**, #1 (2007), pp. 21–30. Bibcode: 2007JAHH...10...21H

[5] " Asteroid (http://encarta.msn.com/encyclopedia_761551567/asteroid.html)", *MSN Encarta*, Microsoft. Accessed May 5, 2008. Archived (http://www.webcitation.org/5kx3nDxZi) 2009-11-01.

[6] Press release, IAU 2006 General Assembly: Result of the IAU Resolution votes (http://www.iau.org/public_press/news/detail/iau0603/), International Astronomical Union, August 24, 2006. Accessed May 5, 2008.

[7] Questions and Answers on Planets (http://www.iau.org/public_press/news/release/iau0603/questions_answers/), additional information, news release IAU0603, IAU 2006 General Assembly: Result of the IAU Resolution votes, International Astronomical Union, August 24, 2006. Accessed May 8, 2008.

[8] JPL. "How Many Solar System Bodies" (http://ssd.jpl.nasa.gov/?body_count). *JPL Solar System Dynamics*. NASA. . Retrieved 2010-09-27.

[9] "Discovery Circumstances: Numbered Minor Planets (1)-(5000)" (http://www.minorplanetcenter.net/iau/lists/NumberedMPs000001.html). Minor Planet Center. . Retrieved 2011-02-26.

[10] "Discovery Circumstances: Numbered Minor Planets (240001)-(245000)" (http://www.minorplanetcenter.net/iau/lists/NumberedMPs240001.html). Minor Planet Center. . Retrieved 2011-02-26.

[11] Yeomans, Don, "Near-Earth Object groups" (http://neo.jpl.nasa.gov/neo/groups.html), *Near Earth Object Project* (NASA), , retrieved 2011-12-24

[12] Connors, Martin; Wiegert, Paul; Veillet, Christian (July 2011), "Earth's Trojan asteroid", *Nature* **475** (7357): 481–483, Bibcode 2011Natur.475..481C, doi:10.1038/nature10233, PMID 21796207

[13] Trilling, David et al. (October 2007), "DDT observations of five Mars Trojan asteroids", *Spitzer Proposal ID #465*, Bibcode 2007sptz.prop..465T

[14] Horner, J.; Evans, N.W.; Bailey, M. E. (2004). "Simulations of the Population of Centaurs I: The Bulk Statistics". *Monthly Notices of the Royal Astronomical Society* **354** (3): 798–810. arXiv:astro-ph/0407400. Bibcode 2004MNRAS.354..798H. doi:10.1111/j.1365-2966.2004.08240.x.

[15] Neptune trojans, Jupiter trojans (http://www.dtm.ciw.edu/users/sheppard/trojans/)

[16] NASA JPL Small-Body Database Browser on 96747 Crespodasilva (http://ssd.jpl.nasa.gov/sbdb.cgi?sstr=96747+Crespodasilva)

[17] Staff (November 28, 2000). "Lucy Crespo da Silva, 22, a senior, dies in fall" (http://hubblesite.org/newscenter/newsdesk/archive/releases/1991/12/text/). Hubble News Desk. . Retrieved 2008-04-15.

[18] 'valley' being a common abbreviation of 'grain' that would be formally adopted with simplified Chinese characters.

[19] "Division III Commission 15 Physical Study of Comets & Minor Planets" (http://www.iau.org/science/scientific_bodies/commissions/15). International Astronomical Union (IAU). September 29, 2005. . Retrieved 2010-03-22.

[20] "Physical Properties of Asteroids" (http://www.psi.edu/pds/archive/physical.html). .

[21] "The Near-Earth Asteroids Data Base" (http://earn.dlr.de/nea). .

[22] http://www.minorplanetcenter.net/iau/mpc.html

List_of_minor_planets

This is a **list of numbered minor planets** in the Solar System, in numerical order.

As of December 2011 there are 310,376 numbered minor planets, and about as many yet unnumbered. Most are not particularly noteworthy; only some 16,000 minor planets have been given names[1] (the first nameless minor planet being number 3708).[2]

For a short list of notable asteroids, see list of notable asteroids, and for notable minor planets beyond Neptune's orbit, see trans-Neptunian object and list of possible dwarf planets.

Five minor planets have been accepted as dwarf planets by the IAU, and hundreds more are likely to be dwarf planets.

Numbering and naming conventions

After discovery, asteroids generally receive a provisional designation (such as "1989 AC"), then a number (such as 4179), and finally (optionally) a name (such as "Toutatis"), in that order.

In modern times, an asteroid receives a sequential number only after its orbit is precisely known. Asteroids whose orbits are not (yet) precisely known are known by their provisional designation. This rule was not necessarily followed in earlier times, and some asteroids received a number but were subsequently "lost". All of these have now been recovered; the last "lost" numbered asteroid was 719 Albert.

Only after a number is assigned is the asteroid eligible to receive a name. Usually the discoverer has up to 10 years to pick a name; some asteroids remain unnamed. Especially towards the end of the twentieth century, with large-scale automated asteroid discovery programs such as LINEAR, the pace of discoveries has increased so much that it seems likely that the vast majority of minor planets will never receive names.

For the reasons mentioned above, the sequence of numbers only approximately matches the timeline of discovery. In extreme cases, such as "lost" asteroids, there may be a considerable mismatch: for instance the high-numbered 69230 Hermes was originally discovered in 1937, but it was a lost asteroid until 2003. Only after it was rediscovered could its orbit be established and a number assigned.

Index to lists of minor planets

The minor planets are listed in the following:

1–1000	10001–11000	20001–21000	30001–31000	40001–41000
1001–2000	11001–12000	21001–22000	31001–32000	41001–42000
2001–3000	12001–13000	22001–23000	32001–33000	42001–43000
3001–4000	13001–14000	23001–24000	33001–34000	43001–44000
4001–5000	14001–15000	24001–25000	34001–35000	44001–45000
5001–6000	15001–16000	25001–26000	35001–36000	45001–46000
6001–7000	16001–17000	26001–27000	36001–37000	46001–47000
7001–8000	17001–18000	27001–28000	37001–38000	47001–48000
8001–9000	18001–19000	28001–29000	38001–39000	48001–49000
9001–10000	19001–20000	29001–30000	39001–40000	49001–50000

50001–51000	60001–61000	70001–71000	80001–81000	90001–91000
51001–52000	61001–62000	71001–72000	81001–82000	91001–92000

52001–53000	62001–63000	72001–73000	82001–83000	92001–93000
53001–54000	63001–64000	73001–74000	83001–84000	93001–94000
54001–55000	64001–65000	74001–75000	84001–85000	94001–95000
55001–56000	65001–66000	75001–76000	85001–86000	95001–96000
56001–57000	66001–67000	76001–77000	86001–87000	96001–97000
57001–58000	67001–68000	77001–78000	87001–88000	97001–98000
58001–59000	68001–69000	78001–79000	88001–89000	98001–99000
59001–60000	69001–70000	79001–80000	89001–90000	99001–100000

100001–101000	110001–111000	120001–121000	130001–131000	140001–141000
101001–102000	111001–112000	121001–122000	131001–132000	141001–142000
102001–103000	112001–113000	122001–123000	132001–133000	142001–143000
103001–104000	113001–114000	123001–124000	133001–134000	143001–144000
104001–105000	114001–115000	124001–125000	134001–135000	144001–145000
105001–106000	115001–116000	125001–126000	135001–136000	145001–146000
106001–107000	116001–117000	126001–127000	136001–137000	146001–147000
107001–108000	117001–118000	127001–128000	137001–138000	147001–148000
108001–109000	118001–119000	128001–129000	138001–139000	148001–149000
109001–110000	119001–120000	129001–130000	139001–140000	149001–150000

150001–151000	160001–161000	170001–171000	180001–181000	190001–191000
151001–152000	161001–162000	171001–172000	181001–182000	191001–192000
152001–153000	162001–163000	172001–173000	182001–183000	192001–193000
153001–154000	163001–164000	173001–174000	183001–184000	193001–194000
154001–155000	164001–165000	174001–175000	184001–185000	194001–195000
155001–156000	165001–166000	175001–176000	185001–186000	195001–196000
156001–157000	166001–167000	176001–177000	186001–187000	196001–197000
157001–158000	167001–168000	177001–178000	187001–188000	197001–198000
158001–159000	168001–169000	178001–179000	188001–189000	198001–199000
159001–160000	169001–170000	179001–180000	189001–190000	199001–200000

200001–201000	210001–211000
201001–202000	211001–212000
202001–203000	212001–213000
203001–204000	213001–214000
204001–205000	214001–215000
205001–206000	215001–
206001–207000	
207001–208000	
208001–209000	
209001–210000	

See also

- List of comets
- List of Solar System bodies formerly regarded as planets
- List of trans-Neptunian objects
- Meanings of minor planet names
- Minor planet (for links to articles on particular groups and families, some of which have lists)
- Minor Planet Center
- Minor-planet moon (includes list)
- Pronunciation of asteroid names

By criteria

- List of instrument-resolved minor planets
- List of minor planets named after people
- List of minor planets named after places
- List of notable asteroids

References

[1] "Minor Planet Statistics" (http://www.minorplanetcenter.org/iau/lists/ArchiveStatistics.html). Minor Planet Center. . Retrieved 2010-09-28.

[2] "Discovery Circumstances: Numbered Minor Planets (1)-(5000)" (http://www.minorplanetcenter.org/iau/lists/NumberedMPs000001.html). Minor Planet Center. . Retrieved 2010-09-28.

- MPC Archive Statistics (http://www.minorplanetcenter.org/iau/lists/ArchiveStatistics.html) (amount of observations, orbits and names)
- MPC Discovery Circumstances (http://www.minorplanetcenter.org/iau/lists/NumberedMPs.html) (minor planets by number)

Further reading

- *Dictionary of Minor Planet Names, 5th ed.: Prepared on Behalf of Commission 20 Under the Auspices of the International Astronomical Union*, Lutz D. Schmadel, ISBN 3-540-00238-3
- *The Names of the Minor Planets*, Paul Herget, 1968, OCLC 224288991

External links

- Lists and plots: Minor Planets (http://www.cfa.harvard.edu/iau/lists/MPLists.html)
- NASA Near Earth Object Program (http://neo.jpl.nasa.gov/)
- PDS Asteroid Data Archive (http://www.psi.edu/pds/)
- SBN Small Bodies Data Archive (http://pdssbn.astro.umd.edu/sbnhtml/index.html)
- JPL Minor Planet Database for physical and orbital data (http://ssd.jpl.nasa.gov/sbdb.cgi?sstr=1 Ceres;orb=1;cov=0;log=0#orb) (of any Small Solar System Body or dwarf planet)

Small_Solar_System_body

A **small Solar System body** (SSSB) is an object in the Solar System that is neither a planet nor a dwarf planet, nor a satellite of a planet or dwarf planet:[1]

> All other objects, except satellites, orbiting the Sun shall be referred to collectively as "Small Solar System Bodies" ... These currently include most of the Solar System asteroids, most Trans-Neptunian Objects (TNOs), comets, and other small bodies.[2]

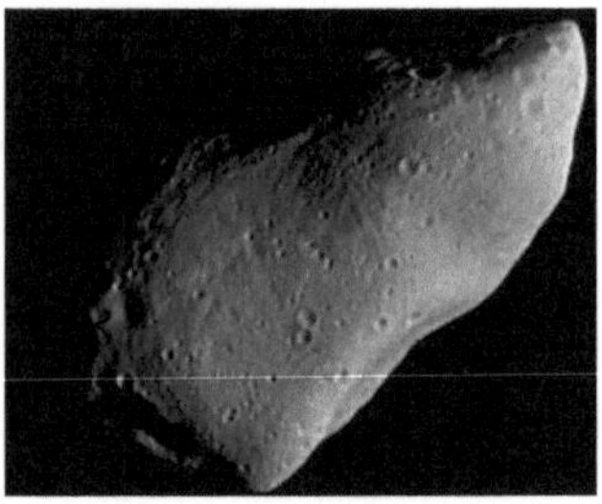

951 Gaspra, a small Solar System body in the asteroid belt, measuring about 18 km in length (photographed by the Galileo probe)

This encompasses all comets and all minor planets other than those classified as dwarf planets, *i.e.*:

- the classical asteroids, with the exception of Ceres;
- the centaurs and trojans;
- the trans-Neptunian objects, with the exception of Pluto, Haumea, Makemake and Eris.

The term was first defined in 2006 by the International Astronomical Union.

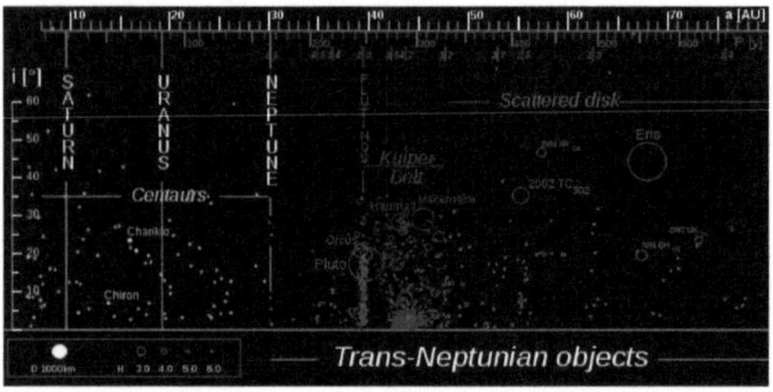

Distribution of centaurs and trans-Neptunian objects

It is not presently clear whether a lower size bound will be established as part of the definition of small Solar System bodies in the future, or if it will encompass all material down to the level of meteoroids, the smallest macroscopic bodies in orbit around the Sun. (On a microscopic level there are even smaller objects such as interplanetary dust, particles of solar wind and free particles of hydrogen.)

Except for the largest, which are in hydrostatic equilibrium, moons differ from small Solar System bodies not in size, but in their orbits. Moons' orbits are not centered around the Sun but around other Solar System objects such as planets, dwarf planets, and even small Solar System bodies themselves.

Some of the larger small Solar System bodies may be reclassified in future as dwarf planets, pending further examination to determine whether or not they are in hydrostatic equilibrium.

The orbits of the vast majority of small Solar System bodies are located in two distinct areas, namely the asteroid belt and the Kuiper belt. These two belts possess some internal structure related to perturbations by the major planets (particularly Jupiter and Neptune, respectively), and have fairly loosely defined boundaries. Other areas of the Solar System also encompass small bodies in smaller concentrations. These include the near-Earth asteroids, centaurs, comets, and scattered disc objects.

See also

- List of Solar System objects by size
- Apollo asteroid
- Asteroid belt
- Centaur (minor planet)
- Comet
- Cybele asteroid
- Hilda family
- Hungaria family
- Kuiper belt
- Lists of small Solar System bodies
- List of gravitationally rounded objects of the Solar System
- List of dwarf-planet candidates
- Meteoroid
- Near-Earth asteroid
- Trojan asteroid
- Trans-Neptunian object
- Vulcanoid asteroid

Notes

[1] The formally correct typography for common nouns such as "small Solar System bodies" and "trans-Neptunian objects" is sentence case, rather than title case as used by the IAU in this instance.

[2] *RESOLUTION B5 - Definition of a Planet in the Solar System* (http://www.iau.org/static/resolutions/Resolution_GA26-5-6.pdf) (IAU)

References

External links

- Expected Science Return of Spatially-Extended In-Situ Exploration at Small Solar System Bodies (http://www.stanford.edu/~pavone/papers/Castillo.Pavone.ea.Aero12.pdf)

Orbital_eccentricity

The **orbital eccentricity** of an astronomical object is a parameter that determines the amount by which its orbit around another body deviates from a perfect circle. A value of 0 is a circular orbit, values between 0 and 1 form an elliptical orbit, 1 is a parabolic escape orbit, and greater than 1 is a hyperbola. The term derives its name from the parameters of conic sections, as every Kepler orbit is a conic section. It is normally used for the isolated two-body problem, but extensions exist for objects following a rosette orbit through the galaxy.

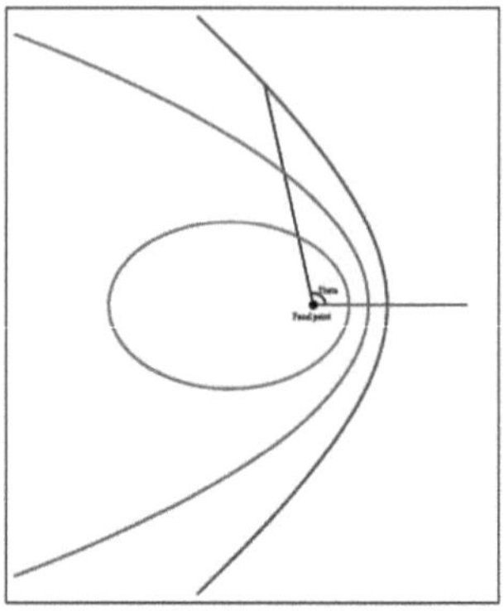

An elliptic Kepler orbit with an eccentricity of 0.7 (red ellipse), a parabolic Kepler orbit (green) and a hyperbolic Kepler orbit with an eccentricity of 1.3 (blue outer line)

Definition

In a two-body problem with inverse-square-law force, every orbit is a Kepler orbit. The eccentricity of this Kepler orbit is a positive number that defines its shape.

The eccentricity may take the following values:

- circular orbit:
- elliptic orbit: (see Ellipse)
- parabolic trajectory: (see Parabola)
- hyperbolic trajectory: (see Hyperbola)

The eccentricity is given by

where *E* is the total orbital energy, is the angular momentum, is the reduced mass. and the coefficient of the inverse-square law central force such as gravity or electrostatics in classical physics:

(is negative for an attractive force, positive for a repulsive one) (see also Kepler problem).

or in the case of a gravitational force:

where is the specific orbital energy (total energy divided by the reduced mass), the standard gravitational parameter based on the total mass, and the specific relative angular momentum (angular momentum divided by the reduced mass).

For values of *e* from 0 to 1 the orbit's shape is an increasingly elongated (or flatter) ellipse; for values of *e* from 1 to infinity the orbit is a hyperbola branch making a total turn of 2 arccsc *e*, decreasing from 180 to 0 degrees. The limit case between an ellipse and a hyperbola, when *e* equals 1, is parabola.

Radial trajectories are classified as elliptic, parabolic, or hyperbolic based on the energy of the orbit, not the eccentricity. Radial orbits have zero angular momentum and hence eccentricity equal to one. Keeping the energy constant and reducing the angular momentum, elliptic, parabolic, and hyperbolic orbits each tend to the corresponding type of radial trajectory while *e* tends to 1 (or in the parabolic case, remains 1).

For a repulsive force only the hyperbolic trajectory, including the radial version, is applicable.

For elliptical orbits, a simple proof shows that arcsin() yields the projection angle of a perfect circle to an ellipse of eccentricity . For example, to view the eccentricity of the planet Mercury (=0.2056), one must simply calculate the inverse sine to find the projection angle of 11.86 degrees. Next, tilt any circular object (such as a coffee mug viewed from the top) by that angle and the apparent ellipse projected to your eye will be of that same eccentricity.

Etymology

From Medieval Latin *eccentricus*, derived from Greek *ekkentros* "out of the center", from *ek-*, *ex-* "out of" + *kentron* "center". Eccentric first appeared in English in 1551, with the definition "a circle in which the earth, sun. etc. deviates from its center." Five years later, in 1556, an adjective form of the word was added.

Calculation

Eccentricity of an orbit can be calculated from orbital state vectors as a magnitude of eccentricity vector:

where:

- is eccentricity vector.

For elliptical orbits it can also be calculated from distance at apoapsis and periapsis:

where:

- is radius at apoapsis (i.e., the farthest distance of the orbit to the center of mass of the system, which is a focus of the ellipse).
- is radius at periapsis (the closest distance).

The eccentricity of an elliptical orbit can also be used to obtain the ratio of the periapsis to the apoapsis:

Examples

The eccentricity of the Earth's orbit is currently about 0.0167; the Earth's orbit is nearly circular. Over hundreds of thousands of years, the eccentricity of the Earth's orbit varies from nearly 0.0034 to almost 0.058 as a result of gravitational attractions among the planets (see graph [2]).[3]

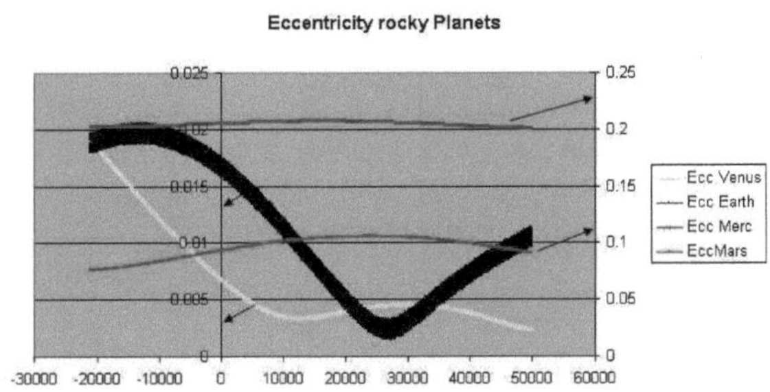

Gravity Simulator [1] plot of the changing orbital eccentricity of Mercury, Venus, Earth, and Mars over the next 50,000 years. The 0 point on this plot is the year 2007.

Mercury has the greatest orbital eccentricity of any planet in the Solar System (e=0.2056). Before 2006, Pluto was considered to be the planet with the most eccentric orbit (e=0.248). The Moon's value is 0.0549. For the values for all planets and other celestial bodies in one table, see *List of gravitationally rounded objects of the Solar System.*

Most of the Solar System's asteroids have orbital eccentricities between 0 and 0.35 with an average value of 0.17.[4] Their comparatively high eccentricities are probably due to the influence of Jupiter and to past collisions.

The eccentricity of comets is most often close to 1. Periodic comets have highly eccentric elliptical orbits with eccentricities just below 1; Halley's Comet's elliptical orbit, for example, has a value of 0.967. Non-periodic comets follow near-parabolic orbits and thus have eccentricities even closer to 1. Examples include Comet Hale–Bopp with a value of 0.995[5] and comet C/2006 P1 (McNaught) with a value of 1.000019.[6] As Hale–Bopp's value is less than 1, its orbit is elliptical and will in fact return.[5] Comet McNaught has a hyperbolic orbit while within the influence of the planets, but is still bound to the Sun with an orbital period of about 10^5 years.[7] As of a 2010 Epoch, Comet C/1980 E1 has the largest eccentricity of any known hyperbolic comet with an eccentricity of 1.057,[8] and will leave the Solar System indefinitely.

Neptune's largest moon Triton has an eccentricity of 1.6×10^{-5},[9] the smallest eccentricity of any known body in the Solar System; its orbit is as close to a perfect circle as can be currently measured.

Mean eccentricity

The mean eccentricity of an object is the average eccentricity as a result of perturbations over a given time period. Neptune currently has an instant (current Epoch) eccentricity of 0.0113,[10] but from 1800 A.D. to 2050 A.D. has a *mean* eccentricity of 0.00859.[11]

Climatic effect

Orbital mechanics require that the duration of the seasons be proportional to the area of the Earth's orbit swept between the solstices and equinoxes, so when the orbital eccentricity is extreme, the seasons that occur on the far side of the orbit (aphelion) can be substantially longer in duration. Today, northern hemisphere fall and winter occur at closest approach (perihelion), when the earth is moving at its maximum velocity. As a result, in the northern hemisphere, fall and winter are slightly shorter than spring and summer. In 2006, summer was 4.66 days longer than winter and spring was 2.9 days longer than fall.[12] Apsidal precession slowly changes the place in the Earth's orbit where the solstices and equinoxes occur (this is not the precession of the axis). Over the next 10,000 years, northern hemisphere winters will become gradually longer and summers will become shorter. Any cooling effect, however, will be counteracted by the fact that the eccentricity of Earth's orbit will be almost halved, reducing the mean orbital radius and raising temperatures in both hemispheres closer to the mid-interglacial peak.

See also

- Eccentricity (mathematics)
- Eccentricity vector
- Equation of time
- Milankovitch cycles
- Orbits

References

[1] http://www.orbitsimulator.com/gravity/articles/what.html

[2] http://www.museum.state.il.us/exhibits/ice_ages/eccentricity_graph.html

[3] A. Berger and M.F. Loutre (1991 (old, but published)). "Graph of the eccentricity of the Earth's orbit" (http://www.museum.state.il.us/exhibits/ice_ages/eccentricity_graph.html). Illinois State Museum (Insolation values for the climate of the last 10 million years). . Retrieved 2009-12-17.

[4] Asteroids (http://filer.case.edu/sjr16/advanced/asteroid.html)

[5] "JPL Small-Body Database Browser: C/1995 O1 (Hale-Bopp)" (http://ssd.jpl.nasa.gov/sbdb.cgi?sstr=Hale-Bopp). 2007-10-22 last obs. . Retrieved 2008-12-05.

[6] "JPL Small-Body Database Browser: C/2006 P1 (McNaught)" (http://ssd.jpl.nasa.gov/sbdb.cgi?sstr=C/2006+P1). 2007-07-11 last obs. . Retrieved 2009-12-17.

[7] "Comet C/2006 P1 (McNaught) - facts and figures" (http://www.perthobservatory.wa.gov.au/information/comet_mcnaught_info.html). Perth Observatory in Australia. 2007-01-22. . Retrieved 2011-02-01.

[8] "JPL Small-Body Database Browser: C/1980 E1 (Bowell)" (http://ssd.jpl.nasa.gov/sbdb.cgi?sstr=1980E1). 1986-12-02 last obs. . Retrieved 2010-03-22.

[9] David R. Williams (22 January 2008). "Neptunian Satellite Fact Sheet" (http://nssdc.gsfc.nasa.gov/planetary/factsheet/neptuniansatfact.html). NASA. . Retrieved 2009-12-17.

[10] Williams, David R. (2007-11-29). "Neptune Fact Sheet" (http://nssdc.gsfc.nasa.gov/planetary/factsheet/neptunefact.html). NASA. . Retrieved 2009-12-17.

[11] "Keplerian elements for 1800 A.D. to 2050 A.D." (http://ssd.jpl.nasa.gov/txt/p_elem_t1.txt). JPL Solar System Dynamics (http://ssd.jpl.nasa.gov/?planet_pos). . Retrieved 2009-12-17.

[12] This information is concerning the summer of the year 2006 not the current year we are in now.

Prussing, John E., and Bruce A. Conway. Orbital Mechanics. New York: Oxford University Press, 1993.

External links

- World of Physics: Eccentricity (http://scienceworld.wolfram.com/physics/Eccentricity.html)
- The NOAA page on Climate Forcing Data (http://www.ncdc.noaa.gov/paleo/forcing.html) includes (calculated) data from Berger (1978), Berger and Loutre (1991) (ftp://ftp.ncdc.noaa.gov/pub/data/paleo/insolation/). Laskar et al. (2004) (http://www.imcce.fr/Equipes/ASD/insola/earth/earth.html) on Earth orbital variations, Includes eccentricity over the last 50 million years and for the coming 20 million years.
- The orbital simulations by Varadi, Ghil and Runnegar (2003) (http://astrobiology.ucla.edu/OTHER/SSO/) provides series for Earth orbital eccentricity and orbital inclination.
- Kepler's Second law's simulation (http://www.walter-fendt.de/ph14e/keplerlaw2.htm)

Orbital_period

The **orbital period** is the time taken for a given object to make one complete orbit about another object.

When mentioned without further qualification in astronomy this refers to the **sidereal period** of an astronomical object, which is calculated with respect to the stars.

There are several kinds of orbital periods for objects around the Sun (or other celestial objects):

- The **sidereal period** is the temporal cycle that it takes an object to make a full orbit, relative to the stars. This is considered to be an object's true orbital period.
- The **synodic period** is the temporal interval that it takes for an object to reappear at the same point in relation to two or more other objects, e.g., when the Moon relative to the Sun as observed from Earth returns to the same illumination phase. The synodic period is the time that elapses between two successive conjunctions with the Sun–Earth line in the same linear order. The synodic period differs from the sidereal period due to the Earth's orbiting around the Sun.
- The **draconitic period**, or **draconic period**, is the time that elapses between two passages of the object through its ascending node, the point of its orbit where it crosses the ecliptic from the southern to the northern hemisphere. This period differs from the sidereal period because both the orbital plane of the object and the plane of the ecliptic precess with respect to the fixed stars, so their intersection, the line of nodes, also precesses with respect to the fixed stars. Although the plane of the ecliptic is often held fixed at the position it occupied at a specific epoch, the orbital plane of the object still precesses causing the draconitic period to differ from the sidereal period.
- The **anomalistic period** is the time that elapses between two passages of an object at its periapsis (in the case of the planets in the solar system, called the perihelion), the point of its closest approach to the attracting body. It differs from the sidereal period because the object's semimajor axis typically advances slowly.
- Also, the Earth's **tropical period** (or simply its "year") is the time that elapses between two alignments of its axis of rotation with the Sun, also viewed as two passages of the object at right ascension zero. One Earth year has a slightly shorter interval than the solar orbit (sidereal period) because the inclined axis and equatorial plane slowly precesses (rotates in sidereal terms), realigning before orbit completes with an interval equal to the inverse of the precession cycle (about 25,770 years).

Relation between the sidereal and synodic periods

Table of synodic periods in the Solar System, relative to Earth:

	Sidereal Period (a)	**Synodic Period** (a)	**Synodic Period** (d)
Solar surface	0.069[1] (25.3 days)	0.074	27.3
Mercury	0.240846 (87.9691 days)	0.317	115.88
Venus	0.615 (225 days)	1.599	583.9
Earth	**1** (365.25636 solar days)	—	—
Moon	0.0748	0.0809	29.5306
Mars	1.881	2.135	779.9
4 Vesta	3.629	1.380	504.0
1 Ceres	4.600	1.278	466.7
10 Hygiea	5.557	1.219	445.4
Jupiter	11.86	1.092	398.9
Saturn	29.46	1.035	378.1
Uranus	84.32	1.012	369.7
Neptune	164.8	1.006	367.5
134340 Pluto	248.1	1.004	366.7
136199 Eris	557	1.002	365.9
90377 Sedna	12050	1.00001	365.1

In the case of a planet's moon, the synodic period usually means the Sun-synodic period. That is to say, the time it takes the moon to complete its illumination phases, completing the solar phases for an observer on the planet's surface —the Earth's motion does not determine this value for other planets, because an Earth observer is not orbited by the moons in question. For example, Deimos' synodic period is 1.2648 days, 0.18% longer than Deimos' sidereal period of 1.2624 d.

Calculation

Small body orbiting a central body

According to Kepler's Third Law, the **orbital period** (in seconds) of two bodies orbiting each other in a circular or elliptic orbit is:

where:

- is the orbit's semi-major axis, in meters
- is the standard gravitational parameter, in
- is the gravitational constant,
- is the mass of the more massive body.

For all ellipses with a given semi-major axis the orbital period is the same, regardless of eccentricity.

Orbital period as a function of central body's density

When a very small body is in a circular orbit barely above the surface of a sphere of radius R and average density ρ in g/cm^3, the above equation simplifies to:

So, for the Earth as central body (or for any other spherically symmetric body with the same average density) we get:

hours

and for a body made of water

hours

Thus, as an alternative for using a very small number like *G*, the strength of universal gravity can be described using some reference material, like water: the orbital period for an orbit just above the surface of a spherical body of water is 3 hours and 18 minutes. Conversely, this can be used as a kind of "universal" unit of time if we have a unit of mass, a unit of length and a unit of density.

Two bodies orbiting each other

In celestial mechanics when both orbiting bodies' masses have to be taken into account the **orbital period** can be calculated as follows:

where:

- is the sum of the semi-major axes of the ellipses in which the centers of the bodies move, or equivalently, the semi-major axis of the ellipse in which one body moves, in the frame of reference with the other body at the origin (which is equal to their constant separation for circular orbits),
- is the sum of the masses of the two bodies,
- is the gravitational constant.

Note that the orbital period is independent of size: for a scale model it would be the same, when densities are the same (see also Orbit#Scaling in gravity).

In a parabolic or hyperbolic trajectory the motion is not periodic, and the duration of the full trajectory is infinite.

Earth orbits

orbit	center-to-center distance	altitude above the Earth's surface	speed	Orbital period	specific orbital energy
Earth's surface (for comparison)	6,400 km	0 km	7.89 km/s (17,650 mph)	—	-62.6 MJ/kg
Low Earth orbit	6,600 to 8,400 km	200 to 2,000 km	circular orbit: 7.8 to 6.9 km/s (17,450 mph to 15,430 mph) respectively elliptic orbit: 8.2 to 6.5 km/s respectively	89 to 128 min	-17.0 MJ/kg
Molniya orbit	6,900 to 46,300 km	500 to 39,900 km	10.0 to 1.5 km/s (22,370 mph to 3,335 mph) respectively	11 h 58 min	-4.7 MJ/kg
GEO	42,000 km	35,786 km	3.1 km/s (6,935 mph)	23 h 56 min	-4.6 MJ/kg
Orbit of the Moon	363,000 to 406,000 km	357,000 to 399,000 km	1.08 to 0.97 km/s (2,416 to 2,170 mph) respectively	27.3 days	-0.5 MJ/kg

Binary stars

Binary star	Orbital period
AM Canum Venaticorum	17.146 minutes
Beta Lyrae AB	12.9075 days
Alpha Centauri AB	79.91 years
Proxima Centauri - Alpha Centauri AB	500,000 years or more

See also

- Geosynchronous orbit derivation
- Sidereal time
- Sidereal year
- Opposition (astronomy)
- List of periodic comets

Notes

[1] The motion of the solar surface is not purely gravitational and therefore does not follow Kepler's laws of motion

Mean_anomaly

In celestial mechanics, **mean anomaly** is a parameter relating position and time for a body moving in a Kepler orbit. It is based on the fact that equal areas are swept in equal intervals of time by a line joining the focus and the orbiting body (Kepler's second law).

The mean anomaly increases uniformly from 0 to radians during each orbit. However, it is not an angle. Due to Kepler's second law, the mean anomaly is proportional to the area swept by the focus-to-body line since the last periapsis.

The mean anomaly is usually denoted by the letter , and is given by the formula:

where n is the mean motion, a is the length of the orbit's semi-major axis, and m are the orbiting masses, and G is the gravitational constant.

The mean anomaly is the time since the last periapsis multiplied by the mean motion, and the mean motion is divided by the duration of a full orbit.

The mean anomaly is one of three angular parameters ("anomalies") that define a position along an orbit; the other two being the eccentric anomaly and the true anomaly. If the mean anomaly is known at any given instant, it can be calculated at any later (or prior) instant by simply adding (or subtracting) where represents the time difference. The other anomalies can hence be calculated.

Formulas

The mean anomaly *M* can be computed from the eccentric anomaly *E* and the eccentricity *e* with Kepler's Equation:

To find the position of the object in an elliptic Kepler orbit at a given time *t*, the mean anomaly is found by multiplying the time and the mean motion, then it is used to find the eccentric anomaly by solving Kepler's equation.

It is also frequently seen:

,

Again *n* is the mean motion. However, *t*, in this instance, is the *time since epoch*, which is how much time has passed since the measurement of M_0 was taken. The value M_0 denotes the *mean anomaly at epoch*, which is the mean anomaly at the time the measurement was taken.

See also

- Kepler orbit
- Ellipse
- Eccentric anomaly
- True anomaly

References

- Murray, C. D. & Dermott, S. F. 1999, *Solar System Dynamics*, Cambridge University Press, Cambridge.
- Plummer, H.C., 1960, *An Introductory treatise on Dynamical Astronomy*, Dover Publications, New York. (Reprint of the 1918 Cambridge University Press edition.)

Orbital_inclination

Inclination in general is the angle between a reference plane and another plane or axis of direction.

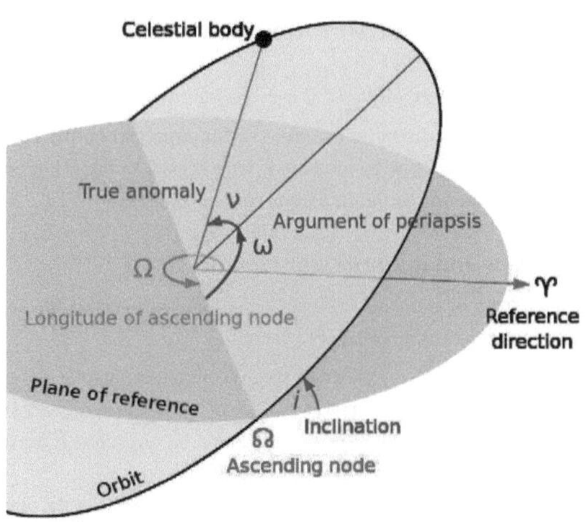

Fig. 1: One view of inclination *i* (green) and other orbital parameters

Orbits

The inclination is one of the six orbital parameters describing the shape and orientation of a celestial orbit. It is the angular distance of the orbital plane from the plane of reference (usually the primary's equator or the ecliptic), normally stated in degrees.[1]

In the Solar System, the inclination of the orbit of a planet is defined as the angle between the plane of the orbit of the planet and the ecliptic — which is the plane containing Earth's orbital path.[2] It could be measured with respect to another plane, such as the Sun's equator or even Jupiter's orbital plane, but the ecliptic is more practical for Earth-bound observers. Most planetary orbits in the Solar System have relatively small inclinations, both in relation to each other and to the Sun's equator. There are notable exceptions in the dwarf planets Pluto and Eris, which have inclinations to the ecliptic of 17 degrees and 44 degrees respectively, and the large asteroid Pallas, which is inclined at 34 degrees.

Inclination				
	Name	**Inclination to ecliptic**	**Inclination to Sun's equator**	**Inclination to invariable plane**[3]
Terrestrials	Mercury	7.01°	3.38°	6.34°
	Venus	3.39°	3.86°	2.19°
	Earth	0°	7.155°	1.57°
	Mars	1.85°	5.65°	1.67°
Gas giants	Jupiter	1.31°	6.09°	0.32°
	Saturn	2.49°	5.51°	0.93°
	Uranus	0.77°	6.48°	1.02°
	Neptune	1.77°	6.43°	0.72°

Natural and artificial satellites

The inclination of orbits of natural or artificial satellites is measured relative to the equatorial plane of the body they orbit if they do so close enough. The equatorial plane is the plane perpendicular to the axis of rotation of the central body.

- an inclination of 0 degrees means the orbiting body orbits the planet in its equatorial plane, in the same direction as the planet rotates;
- an inclination greater than -90° and less than 90° is a prograde orbit.
- an inclination greater than 90° and less than 270° is a retrograde orbit.
- an inclination of exactly 90° is a polar orbit, in which the spacecraft passes over the north and south poles of the planet; and
- an inclination of exactly 180° is a retrograde equatorial orbit.

For the Moon, measuring its inclination with respect to Earth's equatorial plane leads to a rapidly varying quantity and it makes more sense to measure it with respect to the ecliptic (*i.e.* the plane of the orbit that Earth and Moon track together around the Sun), a fairly constant quantity.

Exoplanets and multiple star systems

The inclination of exoplanets or members of multiple stars is the angle of the plane of the orbit relative to the plane perpendicular to the line-of-sight to the object.

- an inclination of 0° is a prograde face-on orbit, meaning the plane of its orbit is parallel to the sky;
- an inclination greater than 0° and less than 90° orbits in the same direction as the rotation of its star;
- an inclination of exactly 90° is an edge-on orbit regardless if the orbit is prograde or retrograde, meaning the plane of its orbit is perpendicular to the sky;
- an inclination greater than 90° and less than 180° orbits in the opposite direction as the rotation of its star; and
- an inclination of 180° is a retrograde face-on orbit, meaning the plane of its orbit is parallel to the sky.

Because the radial velocity method is easier to find planets with more edge-on orbits, most exoplanets would have inclinations between 45° and 135°, even though most exoplanets don't have known inclination. Correspondingly, most exoplanets would have true masses no more than 70% greater than their minimum masses. If the orbit is almost edge-on, then the planet can be seen transiting its star. If the orbit is almost face-on, especially for superjovians detected by radial velocity, then those objects may actually be brown dwarfs or even red dwarfs. One particular example is HD 33636 B, which has true mass 142 M_J, corresponding to an M6V star, while its minimum mass was 9.28 M_J. The inclinations and hence true masses for almost all the exoplanets will eventually be measured by the number of observatories in space, including the Gaia mission, Space Interferometry Mission, and James Webb Space Telescope.

Other meanings

- For planets and other rotating celestial bodies, the angle of the axis of rotation with respect to the normal to plane of the orbit is sometimes also called inclination or axial inclination, but to avoid ambiguity can be called axial tilt or **obliquity**.
- In geology, the magnetic inclination is the angle made by a compass needle with respect to the horizontal surface of the Earth at a given latitude.

Calculation

In astrodynamics, the inclination can be computed from the orbital momentum vector (or any vector perpendicular to the orbital plane) as , where is the z-component of .

Mutual inclination of two orbits may be calculated from their inclinations to another plane using cosine rule for angles.

See also

- Altitude (astronomy)
- Axial tilt
- Azimuth
- Beta Angle
- Kepler orbits
- Kozai effect
- Orbital inclination change

References

[1] Chobotov, Vladimir A. (2002). *Orbital Mechanics* (3rd ed.). AIAA. pp. 28–30;. ISBN 1-56347-537-5.

[2] McBride, Neil; Bland, Philip A.; Gilmour, Iain (2004). *An Introduction to the Solar System*. Cambridge University Press. p. 248. ISBN 0-521-54620-6.

[3] "The MeanPlane (Invariable plane) of the Solar System passing through the barycenter" (http://home.surewest.net/kpheider/astro/MeanPlane.gif). 2009-04-03. . Retrieved 2009-04-10. (produced with Solex 10 (http://chemistry.unina.it/~alvitagl/solex/) written by Aldo Vitagliano)

Longitude_of_the_ascending_node

The **longitude of the ascending node** (☊ or Ω) is one of the orbital elements used to specify the orbit of an object in space. It is the angle from a reference direction, called the *origin of longitude*, to the direction of the ascending node, measured in a reference plane.[1] Commonly used reference planes and origins of longitude include:

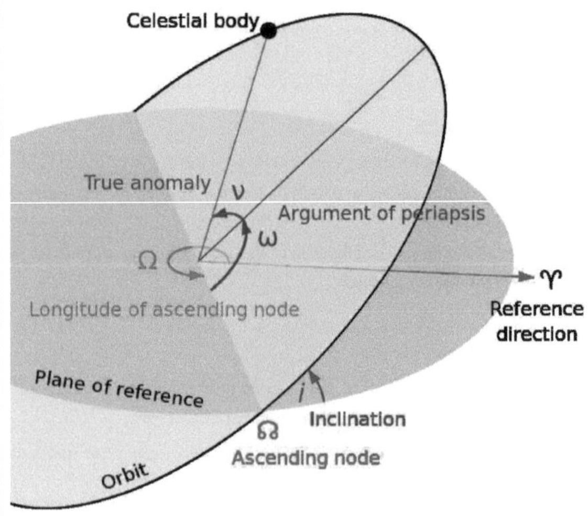

The longitude of the ascending node.

- For a geocentric orbit, Earth's equatorial plane as the reference plane, and the First Point of Aries as the origin of longitude. In this case, the longitude is also called the **right ascension of the ascending node**, or **RAAN**. The angle is measured eastwards (or, as seen from the north, counterclockwise) from the First Point of Aries to the node.[2] [3]
- For a heliocentric orbit, the ecliptic as the reference plane, and the First Point of Aries as the origin of longitude. The angle is measured counterclockwise (as seen from north of the ecliptic) from the First Point of Aries to the node.[2]
- For an orbit outside the Solar System, the plane through the primary perpendicular to a line through the observer and the primary (called the *plane of the sky*) as the reference plane, and north, i.e., the perpendicular projection of the direction from the observer to the North Celestial Pole onto the plane of the sky, as the origin of longitude. The angle is measured eastwards (or, as seen by the observer, counterclockwise) from north to the node.[4] , pp. 40, 72, 137; [5] , chap. 17.

In the case of a binary star known only from visual observations, it is not possible to tell which node is ascending and which is descending. In this case the orbital parameter which is recorded is the **longitude of the node**, Ω, which is the longitude of whichever node has a longitude between 0 and 180 degrees.[5] , chap. 17;[4] , p. 72.

Calculation from state vectors

In astrodynamics, the longitude of the ascending node can be calculated from the specific relative angular momentum vector ***h*** as follows:

Here, $\mathbf{n}=(n_x, n_y, n_z)$ is a vector pointing towards the ascending node. The reference plane is assumed to be the *xy*-plane, and the origin of longitude is taken to be the positive *x*-axis. ***k*** is the unit vector (0, 0, 1), which is the normal vector to the *xy* reference plane.

For non-inclined orbits (with inclination equal to zero), Ω is undefined. For computation it is then, by convention, set equal to zero; that is, the ascending node is placed in the reference direction, which is equivalent to letting ***n*** point towards the positive *x*-axis.

See also

- Kepler orbits
- Equinox
- Orbital node
- perturbation of the orbital plane can cause revolution of the ascending node

References

[1] Parameters Describing Elliptical Orbits (http://www.lns.cornell.edu/~seb/celestia/orbital-parameters.html), web page, accessed May 17, 2007.

[2] Orbital Elements and Astronomical Terms (http://www.physics.ncsu.edu/courses/astron/orbits.html), Robert A. Egler, Dept. of Physics, North Carolina State University. Web page, accessed May 17, 2007.

[3] Keplerian Elements Tutorial (http://www.amsat.org/amsat/keps/kepmodel.html), amsat.org, accessed May 17, 2007.

[4] *The Binary Stars*, R. G. Aitken, New York: Semi-Centennial Publications of the University of California, 1918.

[5] *Celestial Mechanics* (http://astrowww.phys.uvic.ca/~tatum/celmechs.html), Jeremy B. Tatum, on line, accessed May 17, 2007.

Article Sources and Contributors

2695_Christabel *Source*: http://en.wikipedia.org/w/index.php?title=2695_Christabel *Contributors*: Avicennasis, Merovingian, Rich Farmbrough, Scapler, T@nn, Waacstats

Asteroid_belt *Source*: http://en.wikipedia.org/w/index.php?title=Asteroid_belt *Contributors*: 1981willy, 21655, 2T, A-giau, Abbalkea2010, Ace45954, Adam Keller, Aesopos, Ahoerstemeier, Aiken drum, Airplaneman, Akshay Lattimardi, Alansohn, AlexiusHoratius, Alfio, Alsandro, Amara, Anarchist42, Andrewlp1991, Antandrus, Apeman, Aponar Kestrel, Arsia Mons, Art LaPella, Arz1969, Avenue, Avicennasis, AxelBoldt, Baa, Bender235, Bewildebeast, Bfissa, Bigger digger, Blacx0, BlastOButter42, Bobymeltics, Bongwarrior, Bryan Derksen, Burzmali, CanadianLinuxUser, Casliber, CharlesC, Cheat2win, Chesnok, ChristTrekker, ChristopherWillis, CielProfond, CirrusHead, Cjc13, Ckatz, Closedmouth, Coalisnotcools, Colonies Chris, Crash Underride, Crazymonkey1123, DABADABADAI, Dabomb87, Danim, Davewild, Dawnseeker2000, Delrayva, Denisarona, Deon, Deor, Deuar, Discospinster, Donarreiskoffer, Dougano, DrKiernan, Dsparks000, Dspitzle, Dune010, Dwcarless, EarnestyEternity, Eccles999174, Ed Poor, Editore99, Edward130603, Egil, El C, El Cid, Emurphy42, Enviroboy, Epbr123, Everyking, Excirial, Extra999, FF2010, Falcon8765, Fcn, Feens, Felyza, Femto, Fjörgynn, Flyguy649, Folajimi, Fotaun, FrankCarroll, Frenchgizmo, Furrykef, Gaius Cornelius, Gap9551, Gareth Griffith-Jones, Garion96, Geeoharee, Gene Nygaard, Geoff Plourde, George100, Gilderien, Gilliam, Gogo Dodo, Goudzovski, Grahamec, Gtg204y, Gurch, GuyQuest, Gwillhickers, Hadseys, Hairy Dude, HalfShadow, Hamiltonstone, Harry, Hashar, HatriX, Hda3ku, Hdt83, Headbomb, HenryLi, Herostratus, Hibernian, Hiberniantears, Hike395, Hmains, Hope(N Forever), Hunnjazal, I like trains beetches, IVAN3MAN, IanOsgood, Ianare, Icairns, Ilikebakedziti, Improve, Incnis Mrsi, Iridescent, J.delanoy, JForget, JHunterJ, JR00576, Jacj, Jackfork, JamesHoadley, Jaredroberts, Jcronen1, Jeff G., Jeremy Visser, Jim1138, Jivecat, Jnestorius, JocK, Joepearson, John Reaves, John of Reading, JorisvS, Jusdafax, Jusjih, Jwh, Ke6jjj, Kejo13, Kheider, Kintetsubuffalo, Kobrabones, Kristaga, Kuru, Kwamikagami, Kyng, Lakers, Lampman, Lanthanum-138, Latulla, Lavateraguy, Lawrie88f, Leor klier, Lhopitalified, Liamdaly620, Lightmouse, Lilly Ishappy, LonelyMarble, Lotu, Luk, Luokehao, MJ94, Malinadams, Manisero399, Marcus Qwertyus, Marek69, Marksurge, Marskell, Martinwilke1980, Materialscientist, Matthewedwards, Maxamegalon2000, McGeddon, Meelar, Megalodon99, Melara..., Merovingian, Metafax1, Mgiganteus1, Michael Devore, Michael Hardy, Michaelas10, Mike s, Milonica, Moneo, Moocha, Morgan Leigh, MrRadioGuy, Mulad, Murgh, Muriel Gottrop, Murlough23, Mxn, Myanw, Myrrhlin, Narayanese, Nephelin, NerdyScienceDude, Nick C, Nick123, Nivix, NorwegianBlue, Numbers123---, Oldmanjenkins93, Oni Ookami Alfador, Ontyx, Oranjblud, Orii, Ost316, P. S. F. Freitas, P30Carl, Palica, Papa November, Paul H., Pembers, Peyre, Pharos, Philip Trueman, Piano non troppo, Pie4all88, Pinethicket, Pizza Puzzle, Pmanderson, PocketDocket, Poppy, Potatoswatter, PowerCS, Praefectorian, Qu Hongcheng, Quackers074, Quendus, Qurq, RJHall, RP88, RSStockdale, RadicalOne, Ramaksoud2000, Ran, Randomblue, Razorflame, Reaper Eternal, Res2216firestar, Retodon8, Revolución, RexNL, Reywas92, Rhowryn, Rich Farmbrough, Richhoncho, Rjwilmsi, Rmhermen, RobertG, Rogermw, Ronhjones, Rorschach, RossPatterson, Rothorpe, Rst20xx, Ruby.red.roses, Ruslik0, Rwflammang, S Whistler, SERK12, SMSshark, Sardanaphalus, Saros136, Sbig10, SchuminWeb, Scog, ScottyBerg, ScottyBoy900Q, Sdornan, Seaphoto, Sengkang, Serendipodous, Shadowcheets, Shadowjams, Shirik, Shshshsh, Siafu, Sjock, Skizzik, Skysmith, Smartech, SnappingTurtle, Snowolf, Some jerk on the Internet, Something14, Springnuts, Squirepants101, Sry85, Stephen e nelson, Subash.chandran007, Sugarfish, Sumone10154, SuperHamster, SuperSpy00bob, TRUTH316, Taed, Tbhotch, Tcncv, Tearlach, Tedickey, Template namespace initialisation script, Teog, TerraFrost, The High Fin Sperm Whale, The Rambling Man, The Tom, Thehelpfulone, Tide rolls, Tiptoety, Titus III, Tombombadil, Tpbradbury, TravisTX, Treisijs, TrevorX, Tuckerresearch, Ufviper, Uncle Dick, Urhixidur, Vanished User 0001, Viriditas, Visor, Vittsadaf, Vsmith, WRK, Waxcaptain45, Wayne Hardman, Wayne Slam, Weedwhacker128, West-J, Wetman, Whatcanbrowndo, WhiteDragon, Wiki alf, Wiki libs, Wikih101, Wikisux, Wiklol, William Avery, Wimt, Wipsenade, Wolfkeeper, WolfmanSF, Worldtraveller, Wtmitchell, X201, Xoxojoccy, Xxgurlxx, Yosef1987, Zachary, Zamphuor, Zanejune1, Zidane tribal, Zé da Silva, $_0$x, 677 anonymous edits

Asteroid *Source*: http://en.wikipedia.org/w/index.php?title=Asteroid *Contributors*: (, -- April, 10metreh, 129.128.164.xxx, 129.128.90.xxx, 15lsoucy, 1981willy, 1984, 1wonjae, 23skidoo, 28421u2232nfenfcenc, 2help, 3morand, 95j, A-giau, A4, AATTW, ABF, ACSE, AVand, Abc518, Academic Challenger, Acalamari, Access Denied, Accurizer, Adam Keller, Adambro, Addps4cat, AdjustShift, Adnergje, AgadaUrbanit, Ageekgal, Aggie747, Ahoerstemeier, Ajitem49, Ajraddatz, Alan R. Fisher, Alansohn, Ale jrb, Aleenf1, Alex.muller, Alexf, AlexiusHoratius, Alfio, Allstarecho, AllyUnion, Alsandro, Amaury, Amazins490, Amire80, Anaxial, Andonic, Andre Engels, Andrew Gray, AndrewBuck, Andy4789, AngelOfSadness, Angela, AnimAlu, AnkhMorpork, Ann Stouter, Anna512, Antandrus, Antoni Barau, Anville, Apparition11, Apple K, Arcman, Arjun01, Arkuat, Aster2, Asteroidz R not planetz, Astrobiologist, Ato, Attackpig, Avicennasis, Ayelie, Aznrules, Backslash Forwardslash, Badgernet, Banzai314, Bathanyucolkol, Beemer69, Beland, Ben-Zin, Ben.davis35, Benhocking, Bercziszani, Bevo, Bihco, BilCat, Bill37212, Billymac00, Blanchardb, Bluemask, Bob f it, Bobblewik, Bobizuruncle, Bobo192, Bogey97, Bokac1984, Bomac, Bongwarrior, Brandon, Brion VIBBER, Brockert, Bryan Derksen, Bsimmons666, CBDunkerson, CWenger, CWii, Callanecc, Callumw1, Calvin 1998, Can't sleep, clown will eat me, CapitalR, Captain Yesterday, Captain-tucker, Captainbeefart, Carinemily, Carmichael, Carson69, Casmith 789, Catapult, Catttttt2, Celarnor, Ceranthor, Charles Matthews, Charliesface, CharlotteWebb, Chase me ladies, I'm the Cavalry, Chesnok, ChiZeroOne, Chris goulet, Chuck02, Chunkylefunga, Ck lostsword, Ckatz, Ckruschke, Clovis Sangrail, Cmapm, CommonsDelinker, Conversion script, CoolbeansofGR, Coolkid1001, Courcelles, Craigboy, Creidieki, Cryptic, Css, Curps, CzarB, DARTH SIDIOUS 2, DMacks, DVD R W, DW declare, DaL33T, Daddylonglegs420, Dam sens, Danakil, Daniel, DanielCD, Danim, DanwWiki, Daranak, Dartoe, David0811, DavidLevinson, Davidiad, DazBoz, Dcoetzee, Deagle AP, Dekaels, Dendodge, Deor, Deuar, Dffgd, Diderot, Dina, Discospinster, Divebomb, Dman159753, DocWatson42, Dr. Blofeld, DrGatsby1962, Dragon of the Pants, Drbogdan, Dreish, Drift chambers, Drmies, Dstrike224, Duoduoduo, Dwadejames03, Dysepsion, E. Fokker, ERcheck, ESkog, Editor at Large, Eeekster, Eggoeater, Ejosse1, El C, El Shaday, Eleassar, Electron9, Emerson7, Emijrp, Emote, Emurphy42, Enemyofthestate, EoGuy, Epbr123, Eric-Wester, Ericann726, Erik9, Etacar11, EvenGreenerFish, Evil Monkey, Evil saltine, Evvyleeds, Excirial, Eyrian, FM, Fabiform, Fabiob, Faker12345, Falcon8765, Fartherred, Femto, Ferbtastic, Ffbond, Finlay McWalter, Finngall, Firsfron, Fisherama1, Fld300b, Flex Flint, Florian Blaschke, Flyguy649, FlyingToaster, Folajimi, Foobaz, Fosnez, Fountains of Bryn Mawr, Fourohfour, Freakmighty, Freakofnurture, Fredrik, Fritzey97, Fryergreg, Furrykef, Gaius Cornelius, Giftlite, Gilliam, Glacialfox, Glen, Glenn, Globalsolidarity, Gogo Dodo, Golbez, Golgofrinchian, Gomada, GorillaWarfare, Graham87, Greenmango, GregorB, Greswik, Grstain, Gtg204y, Gtrmp, Guanaco, Gurch, Gwillhickers, H1nkles, Hackerrye, Halfway to never, Hamtechperson, Hanacy, Hanberke, Happysailor, Hapsiainen, Hdt83, Helpmet, HenryLi, Hevron1998, Hiddekel, Hiddenfromview, Hike395, Holoeconomics, HorsePunchKid, Hydrogen Iodide, Icairns, Immunize, Insanity Incarnate, Ioscius, Iridescent, Island Monkey, Ispy1981, J 1982, J appleseed2, J.delanoy, JDG, JForget, JFreeman, JHunterJ, JYolkowski, Ja 62, Jacob1207, Jadtnr1, Jake838, Jaksmata, JamieS93, Jantzen is cool, Jared Preston, JavierMC, Jiddisch, Jim.henderson, Jmrowland, Jn1611, Joefromrandb, Joelwest, John Cumbers, John Deming, John254, Jojhutton, Jon Harald Søby, Jonathunder, JorisvS, Jose77, Josh-Levin@ieee.org, Josh70597, Jpk, Jpo, Jschnur, Jusdafax, JustAddPeter, Jwrosenzweig, Jyril, Jóna Þórunn, Kazvorpal, Kborer, Keilana, Kelly Martin, Kenohki, KevWang, Kevyn, Kheider, Kier07, Kiiwiii, King of Hearts, Kingpin13, Kirill Lokshin, Kjoonlee, KleinKlio, Klilidiplomus, Kozawa, Kubigula, Kudret abi, Kurtan, Kuru, Kwamikagami, L337 kybldmstr, LALALA26, La Parka Your Car, Lamb99, Latics, Latitudinarian, Leevanjackson, Lesnail, Lexor, Lifefeed, Lightmouse, Lightyear69, Liken to a glove, LilliPuppy, Limonns, Ling.Nut, Logan, Lolihatepizza, Looxix, Loren.wilton, Loresayer, Lowe4091, Luckybeargod, Lugia2453, Luna Santin, Lupin, MATRIX, MER-C, MONGO, Macedonian, MacroDaemon, Maelnuneb, Magioladitis, Magister Mathematicae, Majorclanger, Mandarax, Mani1, Manmobile, MarcK, Marek69, Mark Foskey, Martarius, Martin451, Mas 18 dl, Masonbarge, Master Jay, Materialscientist, Mathewignash, Mattman00000, Matubo, Maury Markowitz, Mav, MavrickWeirdo, Mbc362, McSly, Mechanical digger, Mejor Los Indios, Menchi, Mendelssohn is Cool, Mephistophelian, Metalb, Metebelis, Mgmirkin, Michael Hardy, Micky kwoskie, Midnightcomm, Mike s, MikeCapone, MikeLynch, Mikenorton, Minesweeper, Minimac, MithrandirAgain, Mjuancarlo, Moomoomoo, Morwen, Moverton, Mrholybrain, Mtnerd, Muijz, Murgh, Murlough23, Mxn, Myanw, Mygerardromance, Myuphrid, NAHID, Nakon, Narcosis17, NawlinWiki, Neier, Nephelin, Nergaal, Netoholic, Newsaholic, Nick UA, Nihiltres, NiiSSA, Niobrara, Njaelkies Lea, Noclevername, Novangelis, Nubiatech, Ocaasi, Oleg Alexandrov, Olegwiki, Omicronpersei8, Orangemarlin, Oranjblud, OwenX, Oxymoron83, P. S. F. Freitas, PTJoshua, PTSE, Patanet7, Patmic2502, Patrick, Paul H., Paul Pot, Pauli133, Pedro, Perey, Pgan002, Phantomsteve, Philip Trueman, Pi@k, Pickom, Pinethicket, Piolinfax, Pit, Pizza Puzzle, PoeticVerse, Polkadotbaby0772, Polyamorph, Ponder, Porterjoh, Powered, Prashanthns, Pseudomonas, Python eggs, Q21op, Q43, Quadell, Quantpole, Quuxplusone, RFerreira, RJHall, RP88, RUL3R, RadicalOne, Rajabhattacharya, Ramesh Ramaiah, RandomCritic, Ravidreams, RazorICE, Razorflame, Rchandra, Reaper Eternal, Red Slash, RedWolf, Res2216firestar, Rettetast, RexNL, Reyk, Rich Farmbrough, Rivajunkie, Rjwilmsi, Rmhermen, RockMFR, Roentgenium111, Ron Ritzman, Rorschach, Rothorpe, Rubicon, Ruslik0, Rwflammang, Ryjaz, SJP, ST47, SU Linguist, Sagark86, Salmar, Sandhuharpreet, Sang'gre Habagat, Sapphic, Sardanaphalus, Saros136, Satori, SaveThePoint, Savemaxim, Saxsux, SchfiftyThree, Schneelocke, SchnitzelMannGreek, Sciurinæ, Sdornan, Senator Palpatine, Serendipodous, Serpentinite, Seth Ilys, Shadowjams, Shalom Yechiel, Shanes, Shirifan, Shotgun5559, Shushruth, SiliconCerebrate, SkerHawx, Skier Dude, Skizzik, SkyWalker, Slicedoranges, Slightsmile, Slon02, Snowolf, Soccer dude 07, Sodium, Sole Soul, Some jerk on the Internet, Something14, Soonhuat95, Southern Illinois SKYWARN, Sp82, Spacepotato, Spartaz, Spartenxiv, Spencer, Spitfire, Standunne, StarryGrandma, Stealingitstrongly, Stemonitis, Stephenchou0722, Steven Zhang, StitchRFC, Stormwriter, Sturmde, Stwalkerster, Subversive.sound, SummerPhD, SuperHamster, Superbeecat, Susan714, Sushi, Svetovid, Swatjester, Synchronism, Tailpig, Tassedethe, Taxman, Tbhotch, Telescopi, Template namespace initialisation script, That Guy, From That Show!, The Enlightened, The High Fin Sperm Whale, The Phoenix, The Singing Badger, The Thing That Should Not Be, TheGWO, TheKoG, Theda, Thestupidguy, ThreeBlindMice, Thue, Tide rolls, TigerShark, Tillwe, Tim123321Tim, Timir2, Timrollpickering, Timwi, Titoxd, Tjkiller555, Tom Peters, TomasBat, Tombomp, Tony Mach, Topgearstopviewer, TransUtopian, Trash boy, Tree Biting Conspiracy, Trekphiler, Tresiden, Tschild, Tyleradkins31, Ultra two, Uni223344, Untifler, Unyoyega, Uppland, Urhixidur, Vansh9988, Vary, Vaĝinator, Vgranucci, Vianello, Voxpuppet, Vrenator, Vsmith, Wackywace, Waltpohl, Waninge, Ward20, Wayne Hardman, Wayward, Weridiot, Werockthescienceclass, Westwood25, Wgungfu, Whitepaw, Why Not A Duck, Wiki alf, Wiki13, Wikiborg, Wikid77, Wikipelli, Wildyoda, Willhsmit, Willshark, WilyD, Wimt, Wipsenade, Wjfox2005, Wknight94, Wolfkeeper, WolfmanSF, Wwheaton, X relentless x, XJamRastafire, Xanucia, Xav71176, Xeno, Xerxes314, Xn44, Yamamoto Ichiro, Yoguys021298, Youknowandy, Yunshui, Z4ngetsu, Zidane tribal, Zsinj, Zundark, Zzzzz, ^demon, Île flottante, Саша Стефановић, Шуруан, 1442 anonymous edits

Edward_L._G._Bowell *Source*: http://en.wikipedia.org/w/index.php?title=Edward_L._G._Bowell *Contributors*: Chesnok, Cuppysfriend, Curps, D6, David Schaich, DerHerrMigo, Docu, Fxcxt, Fylbecatulous, Joefromrandb, John Belushi, M3tainfo, Martarius, Metronomo, Ohconfucius, Petronas, Phil Boswell, Qxz, Seriv, Stanlekub, Szoltys, Template namespace initialisation script, Timrollpickering, Trekphiler, Triddle, Urhixidur, Xyragx, YUL89YYZ, 17 anonymous edits

Anderson_Mesa_Station *Source*: http://en.wikipedia.org/w/index.php?title=Anderson_Mesa_Station *Contributors*: Accurizer, Arpingstone, Backspace, Bahamut0013, Bchampion, Beland, CRKingston, Chachu207, Condem, D6, Eumolpo, Gary King, General Eisenhower, Grutness, Hmains, JCHall, Jake0geek, Klemen Kocjancic, Mailer diablo, Manuel Trujillo Berges, Mike Peel, Mlm42, Nebular110, NekoDaemon, Nono64, RJHall, Sardanaphalus, Saxbryn, Seb az86556, Sgt Pinback, Skapur, Thetrick, Urhixidur, Vegaswikian, 5 anonymous edits

Minor_planet *Source*: http://en.wikipedia.org/w/index.php?title=Minor_planet *Contributors*: 3coma14, 7&6=thirteen, 786b6364, Aaron of Mpls, Aesopos, Alfons2, An Justified Wikipedian, Arsia Mons, Asikal1, Bmclaughlin9, Bryan Derksen, Chesnok, Ckatz, Colinmartin74, David Shepheard, Dicklyon, Eeekster, Fartherred, Fjörgynn, Glenn L, Gpermant, GraL, Interstellar Man,

JHunterJ, Jolielegal, JorisvS, Kayau, Kheider, Kwamikagami, LeaveSleaves, MER-C, Max rspct, Merovingian, Michael Hardy, Miremare, Mrbenten, Murgh, Nergaal, Niobrara, Novangelis, Nubiatech, R'n'B, RJHall, Raymondwinn, Reyk, Rjwilmsi, Rmhermen, Roentgenium111, Romit3, Rorschach, Rothorpe, Ruslik0, Serendipodous, Sfahey, Shirt58, Sobreira, Soulkeeper, Spacepotato, Spidern, Terraflorin, The Tom, The way, the truth, and the light, Thingg, Thorncrag, Tlesher, Trekphiler, Ultra two, Urhixidur, Vağinator, Velpaedia Jenkuklordanus, West.andrew.g, Wikiprodder, Wiklol, William Avery, Willking1979, Woohookitty, WooperJeff, Writtenright, Wwannsda, 80 anonymous edits

List_of_minor_planets *Source*: http://en.wikipedia.org/w/index.php?title=List_of_minor_planets *Contributors*: 84user, Ace45954, Adrian.benko, Aesopos, Ann Stouter, Babur, Beland, Bissinger, Bryan Derksen, Bwithh, Cerejota, Chesnok, Chudacoff, Colonies Chris, Cromwellt, Cureden, Curps, Cyde, Danakil, Danim, DaveGorman, DocWatson42, Dougweller, Enirac Sum, Epbr123, Filiosus's Saga, Fotaun, Giftlite, Gjd001, GregorB, Habj, Ilvon, Imzogelmo, JPG-GR, John fitzburn, John254, Johnnyb2000, Jonathunder, JorisvS, Joseph Dwayne, Jyril, Kheider, KingofGlory, Kwamikagami, LarryGilbert, Li-sung, Lowe4091, Lzur, Merovingian, MindscapesGraphicDesign, Mosesofmason, Nergaal, Njardarlogar, Nuno Tavares, Oerjan, Oliphaunt, Ordishj, Ostracod, Owen, Peteforsyth, Phil Boswell, Piano non troppo, Pooet, Post Falls Man, Python eggs, RJHall, Rbraunwa, Retaggio, Rich Farmbrough, Rmhermen, Roentgenium111, Rory096, Rothorpe, S Marshall, SD6-Agent, Scwlong, Sdrtirs, Serendipodous, Seth Ilys, Shalom Yechiel, Tavix, The Singing Badger, The Tom, Torchiest, Truthanado, U0000, Urhixidur, Wer900, West-J, Wiklol, WilyD, XJamRastafire, Yctaabpjic, Zanimum, 53 anonymous edits

Small_Solar_System_body *Source*: http://en.wikipedia.org/w/index.php?title=Small_Solar_System_body *Contributors*: Alexandrasolender, Andrew Gray, Andrew William Moerbeke, Artwine, Asikal1, Bluap, Ckatz, Cogito ergo sumo, Crzer07, Ctachme, Danny, David Kernow, Deuar, Dicklyon, Dr. Submillimeter, Dreg743, Dude1818, FairHair, Favonian, Ffudgebucket, Filemon, Filiosus's Saga, Fotaun, Geboy, GwydionM, Hrmanu, Imrek, Interstellar Man, Joeblakesley, JorisvS, Joseph Solis in Australia, Jrockley, Jyril, Kwamikagami, Lunokhod, Markov, Materialscientist, Mollwollfumble, Negadrive, NellieBly, Nergaal, Njardarlogar, Novangelis, Ohfosho, Orora, Paddu, Prvc, RandomCritic, RekishiEJ, RingManX, Rmhermen, Robert Weemeyer, Rothorpe, Ruslik0, Serpentinite, Shadowmorph, Sohelpme, Some jerk on the Internet, Something14, Tarotcards, The Enlightened, TheAllSeeingEye, TheTross, Tktktk, Tom Peters, UBIPETRUS, Urhixidur, Vegaswikian, Williams0319, WolfmanSF, عبادة مجاهد ديرانية, 77 anonymous edits

Orbital_eccentricity *Source*: http://en.wikipedia.org/w/index.php?title=Orbital_eccentricity *Contributors*: 0.39, 26 Ramadan, A5b, After Shock, Alec Connors, Almuhammedi, Atlan, Bartsas, Beanyk, BobKawanaka, Caerwine, Callipides, ChrisfromHouston, Ckatz, CosineKitty, DerHexer, Double sharp, Douglas Spencer, Dougofborg, Driftwood87, ELApro, El C, Erik J, Erkcan, EstebanF, Fjörgynn, Geoffrey.landis, Gogo Dodo, GregBenson, Gurchzilla, Hairy Dude, Harley peters, Hyunrosa91, I Like Cheeseburgers, JakeVortex, JasonKnade, Jayy357, Jfmantis, JorisvS, Jyril, Keenan Pepper, Kheider, King Hildebrand, Kmarinas86, Konradek, Lambiam, Larryisgood, Ling.Nut, Lvzon, Moneya, Mustafa 03011, NatePhysics, Neitram, Nihiltres, Nimesh Mistry, Novangelis, Octaazacubane, Padurar2009, Palica, Patrick, Pfvlloyd, Pgan002, Philip tao, PianoDan, Pigman, Portalian, RJHall, RazorICE, Renacat, Ricky81682, Rjwilmsi, Rsrikanth05, RuM, ScottAlanHill, Shenme, Simone, Smartech, Smb1001, SpeedyGonsales, Stamcose, Swpb, Syp, Teapeat, Technopat, Tennekis, Test-tools, Vinay Jha, W0lfie, Warren oO, WolfmanSF, Wormholio, Wælgæst wæfre, عبادة مجاهد ديرانية, 128 anonymous edits

Orbital_period *Source*: http://en.wikipedia.org/w/index.php?title=Orbital_period *Contributors*: 0.39, 16@r, ABF, Abyssoft, Arakish, Argo Navis, Aspects, Avenged Eightfold, AxelBoldt, Behco, Bobby D. DS., Bunnyhop11, Cherkash, Clarince63, CosineKitty, Coulatssa, Crazy Ivan, Curps, Dan Gluck, Deuar, EAderhold, Eagleal, Filemon, GeorgeTSLC, Gerrit, Gob Lofa, Grim Revenant, Harley peters, Headbomb, Heimstern, Hkoala, IRP, Icairns, Inquisitus, JDCMAN, Jecowa, Jeff Cross, Joe Kress, John.szucs, JohnCastle, JohnOwens, JorisvS, Karl Palmen, Keith Edkins, Kheider, Kingpin13, Kristaga, Lenko, Lir, Looxix, Maddie!, Michael Hardy, Mmeijeri, N2e, NOrbeck, Nicksh, Nimesh Mistry, Novalled, Oem2s3, Oftopladb, Onebravemonkey, OverQuantum, Paine Ellsworth, Patrick, PedroPVZ, Pendragon5, Pie4all88, Pizza Puzzle, Portalian, RJHall, RayTomes, RuM, Satori, Sir48, Slash, Smartech, Sverdrup, Swpb, Talon Artaine, Teapeat, Timwi, Tomkinsr, Tristanb, Trystykat, U3002, Urhixidur, Vreezkid, White Ash, Witchunter, Wlodzimierz, Wolfkeeper, Wwoods, XJamRastafire, Xionbox, Zambani, Zedshort, Zhatt, ری‌بدرس, 124 anonymous edits

Mean_anomaly *Source*: http://en.wikipedia.org/w/index.php?title=Mean_anomaly *Contributors*: 0.39, Albmont, AndrewBuck, AstroMark, Beland, Chochopk, CiaPan, Cmapm, CosineKitty, FractalFusion, Iyer.arvind.sundaram, Japanese Searobin, Jfmantis, Jorge Stolfi, Karn, Kwantus, Lasunncty, Liberatus, Michael Hardy, Mike s, NOrbeck, Nikitadanilov, Nviladkar, PAR, Ron E, Sax Russell, Smartech, Stamcose, Swpb, Tamfang, Teapeat, Urhixidur, XJamRastafire, ری‌بدرس, 15 anonymous edits

Orbital_inclination *Source*: http://en.wikipedia.org/w/index.php?title=Orbital_inclination *Contributors*: 0.39, 102orion, 16@r, Ageekgal, Aldaron, Alec Connors, AllanManangan, Andattaca2010, Argo Navis, Attilios, Avjoska, B.d.mills, Bjankuloski06en, Bryan Derksen, Double sharp, DouglasHeld, Dreadstar, Eliyak, Enfenion, Firth m, Gits (Neo), Gtstricky, Hans Dunkelberg, Harp, Icairns, JMK, Jantaro, Javier Carro, Jennavecia, JorisvS, Jyril, Kheider, KurtRaschke, Kwamikagami, Lasunncty, Lir, Lkesteloot, MacTire02, Martin Rizzo, MiShogun, Mike s, N2e, Natox, Newone, Novangelis, Oliver Jennrich, PaterMcFly, PedroPVZ, Piet Delport, PlanetStar, Pomona17, Portalian, RJHall, Rodasmith, Sbootth, Smartech, Spencer, Spiral5800, Swpb, Teapeat, Theoprakt, Tom Peters, Trekphiler, Urhixidur, V1adis1av, Valich, WolfmanSF, Zbayz, Zeimusu, ری‌بدرس, 67 anonymous edits

Longitude_of_the_ascending_node *Source*: http://en.wikipedia.org/w/index.php?title=Longitude_of_the_ascending_node *Contributors*: 0.39, Aldaron, Alison, Charles Matthews, Cybercobra, DuncanHill, Easphi, Edsanville, Japanese Searobin, Karl-Henner, Kemperb, Lasunncty, Mejor Los Indios, Mikeharris111, Mirecki, Monkbel, Nihiltres, Oliver Jennrich, Pomona17, Portalian, Poulpy, RJHall, Rememberway, Ronebofh, Sax Russell, Siafu, Simone, Smartech, Spacepotato, Sriram.aeropsn, Swpb, Teapeat, Thiseye, Urhixidur, V1adis1av, 10 anonymous edits

Image Sources, Licenses and Contributors

Image:InnerSolarSystem-en.png *Source*: http://en.wikipedia.org/w/index.php?title=File:InnerSolarSystem-en.png *License*: unknown *Contributors*: Original uploader was Mdf at en.wikipedia

Image:Giuseppe Piazzi.jpg *Source*: http://en.wikipedia.org/w/index.php?title=File:Giuseppe_Piazzi.jpg *License*: unknown *Contributors*: Howcheng, Lalupa, Lotse, Papa November, Rentier, Rootology, Ruslik0

Image:Main belt i vs a.png *Source*: http://en.wikipedia.org/w/index.php?title=File:Main_belt_i_vs_a.png *License*: unknown *Contributors*: Piotr Deuar

Image:951 Gaspra.jpg *Source*: http://en.wikipedia.org/w/index.php?title=File:951_Gaspra.jpg *License*: unknown *Contributors*: Alaniaris, BokicaK, Papa November, Ruslik0, The Evil IP address, Tryphon, Yonidebest, 5 anonymous edits

Image:AllendeMeteorite.jpg *Source*: http://en.wikipedia.org/w/index.php?title=File:AllendeMeteorite.jpg *License*: unknown *Contributors*: User:Vesta

Image:Main belt e vs a.png *Source*: http://en.wikipedia.org/w/index.php?title=File:Main_belt_e_vs_a.png *License*: unknown *Contributors*: Original uploader was Deuar at en.wikipedia

Image:Kirkwood Gaps.svg *Source*: http://en.wikipedia.org/w/index.php?title=File:Kirkwood_Gaps.svg *License*: unknown *Contributors*: based on plot by Alan Chamberlain, JPL/Caltech

Image:zodiacal.jpg *Source*: http://en.wikipedia.org/w/index.php?title=File:Zodiacal.jpg *License*: unknown *Contributors*: Original uploader was Serendipodous at en.wikipedia

Image:Asteroid proper elements i vs e.png *Source*: http://en.wikipedia.org/w/index.php?title=File:Asteroid_proper_elements_i_vs_e.png *License*: unknown *Contributors*: Bryan Derksen, Deuar, Papa November, Ruslik0, WinstonSmith, Yann, 1 anonymous edits

Image:Dawn Flight Configuration 2.jpg *Source*: http://en.wikipedia.org/w/index.php?title=File:Dawn_Flight_Configuration_2.jpg *License*: unknown *Contributors*: GDK

Image:Asteroidsscale.jpg *Source*: http://en.wikipedia.org/w/index.php?title=File:Asteroidsscale.jpg *License*: unknown *Contributors*: NASA/JPL-Caltech/JAXA/ESA

Image:4 Vesta 1 Ceres Moon at 20 km per px.png *Source*: http://en.wikipedia.org/w/index.php?title=File:4_Vesta_1_Ceres_Moon_at_20_km_per_px.png *License*: unknown *Contributors*: NASA, composition by Urhixidur

Image:Ceres symbol.svg *Source*: http://en.wikipedia.org/w/index.php?title=File:Ceres_symbol.svg *License*: unknown *Contributors*: Aquantrum, Crei0, Galphi, Lexicon, Ramiy, Ruslik0, Sarang, Sarilho1, Urhixidur, 4 anonymous edits

Image:Ceres2.svg *Source*: http://en.wikipedia.org/w/index.php?title=File:Ceres2.svg *License*: unknown *Contributors*: Varun Kumar

Image:Ceres3.svg *Source*: http://en.wikipedia.org/w/index.php?title=File:Ceres3.svg *License*: unknown *Contributors*: Varun Kumar

Image:2Pallas symbol.svg *Source*: http://en.wikipedia.org/w/index.php?title=File:2Pallas_symbol.svg *License*: unknown *Contributors*: User:KleinKlio

Image:2 Pallas.svg *Source*: http://en.wikipedia.org/w/index.php?title=File:2_Pallas.svg *License*: unknown *Contributors*: Vectorized by

Image:Juno symbol.svg *Source*: http://en.wikipedia.org/w/index.php?title=File:Juno_symbol.svg *License*: unknown *Contributors*: Lexicon, Sarang, Urhixidur

Image:3 Juno (1).svg *Source*: http://en.wikipedia.org/w/index.php?title=File:3_Juno_(1).svg *License*: unknown *Contributors*: User:Sim0nx

Image:Symbol 3.jpg *Source*: http://en.wikipedia.org/w/index.php?title=File:Symbol_3.jpg *License*: unknown *Contributors*: Bkell, Garion96, Pickom, RP88

Image:Vesta symbol.svg *Source*: http://en.wikipedia.org/w/index.php?title=File:Vesta_symbol.svg *License*: unknown *Contributors*: G.dallorto, Lexicon, Rmhermen, Sarang, Urhixidur

Image:4 Vesta (0).svg *Source*: http://en.wikipedia.org/w/index.php?title=File:4_Vesta_(0).svg *License*: unknown *Contributors*: Roland Geider

Image:Simbolo di Vesta.svg *Source*: http://en.wikipedia.org/w/index.php?title=File:Simbolo_di_Vesta.svg *License*: unknown *Contributors*: Original uploader was Kwamikagami at en.wikipedia

Image:4 Vesta Unsimplified Symbol.svg *Source*: http://en.wikipedia.org/w/index.php?title=File:4_Vesta_Unsimplified_Symbol.svg *License*: unknown *Contributors*: Original uploader was Tooto at en.wikipedia

File:5 Astraea symbol alternate.svg *Source*: http://en.wikipedia.org/w/index.php?title=File:5_Astraea_symbol_alternate.svg *License*: unknown *Contributors*: Wilson, John

Image:5 Astraea Symbol.svg *Source*: http://en.wikipedia.org/w/index.php?title=File:5_Astraea_Symbol.svg *License*: unknown *Contributors*: User:Rmhermen

Image:6 Hebe Astronomical Symbol.svg *Source*: http://en.wikipedia.org/w/index.php?title=File:6_Hebe_Astronomical_Symbol.svg *License*: unknown *Contributors*: User:Rmhermen

Image:7 Iris Astronomical Symbol.svg *Source*: http://en.wikipedia.org/w/index.php?title=File:7_Iris_Astronomical_Symbol.svg *License*: unknown *Contributors*: Editor at Large, Ruslik0, Urhixidur, Wikiborg

Image:8 Flora Astronomical Symbol.svg *Source*: http://en.wikipedia.org/w/index.php?title=File:8_Flora_Astronomical_Symbol.svg *License*: unknown *Contributors*: Editor at Large, Urhixidur

Image:9 Metis symbol.svg *Source*: http://en.wikipedia.org/w/index.php?title=File:9_Metis_symbol.svg *License*: unknown *Contributors*: User:Bryan Derksen

File:10 Hygeia symbol alternate.svg *Source*: http://en.wikipedia.org/w/index.php?title=File:10_Hygeia_symbol_alternate.svg *License*: unknown *Contributors*: Hind, J. R.

Image:10 Hygiea Astronomical Symbol.svg *Source*: http://en.wikipedia.org/w/index.php?title=File:10_Hygiea_Astronomical_Symbol.svg *License*: unknown *Contributors*: User:Editor at Large

File:11 Parthenope symbol alternate.svg *Source*: http://en.wikipedia.org/w/index.php?title=File:11_Parthenope_symbol_alternate.svg *License*: unknown *Contributors*: Wilson, John

Image:11 Parthenope symbol.svg *Source*: http://en.wikipedia.org/w/index.php?title=File:11_Parthenope_symbol.svg *License*: unknown *Contributors*: User:Editor at Large

Image:12 Victoria symbol.svg *Source*: http://en.wikipedia.org/w/index.php?title=File:12_Victoria_symbol.svg *License*: unknown *Contributors*: User:Editor at Large

File:13 Egeria symbol.svg *Source*: http://en.wikipedia.org/w/index.php?title=File:13_Egeria_symbol.svg *License*: unknown *Contributors*: Hind, John Russell

File:14 Irene symbol.svg *Source*: http://en.wikipedia.org/w/index.php?title=File:14_Irene_symbol.svg *License*: unknown *Contributors*: John Wilson

Image:15 Eunomia symbol.svg *Source*: http://en.wikipedia.org/w/index.php?title=File:15_Eunomia_symbol.svg *License*: unknown *Contributors*: User:Bryan Derksen

File:16 Psyche symbol.svg *Source*: http://en.wikipedia.org/w/index.php?title=File:16_Psyche_symbol.svg *License*: unknown *Contributors*: J. R. Hind

Image:17 Thetis symbol.png *Source*: http://en.wikipedia.org/w/index.php?title=File:17_Thetis_symbol.png *License*: unknown *Contributors*: Symbol created by R. Luther

Image:18 Melpomene symbol.svg *Source*: http://en.wikipedia.org/w/index.php?title=File:18_Melpomene_symbol.svg *License*: unknown *Contributors*: Hind, John Russell

File:19 Fortuna symbol.svg *Source*: http://en.wikipedia.org/w/index.php?title=File:19_Fortuna_symbol.svg *License*: unknown *Contributors*: Hind, John Russell (May 12, 1823 – December 23, 1895)

Image:26 Proserpina symbol.svg *Source*: http://en.wikipedia.org/w/index.php?title=File:26_Proserpina_symbol.svg *License*: unknown *Contributors*: Symbol created by R. Luther

Image:28 Bellona symbol.svg *Source*: http://en.wikipedia.org/w/index.php?title=File:28_Bellona_symbol.svg *License*: unknown *Contributors*: User:Editor at Large

Image:29 Amphitrite symbol.svg *Source*: http://en.wikipedia.org/w/index.php?title=File:29_Amphitrite_symbol.svg *License*: unknown *Contributors*: Chauncey A. Goodrich and Noah Porter, editors

Image:35 Leukothea symbol.png *Source*: http://en.wikipedia.org/w/index.php?title=File:35_Leukothea_symbol.png *License*: unknown *Contributors*: G. Rumker and C.A.F. Peters

Image:37 Fides symbol.svg *Source*: http://en.wikipedia.org/w/index.php?title=File:37_Fides_symbol.svg *License*: unknown *Contributors*: R. Luther

Image:243 ida.jpg *Source*: http://en.wikipedia.org/w/index.php?title=File:243_ida.jpg *License*: unknown *Contributors*: NASA/JPL

Image:Asteroid 2004 FH.gif *Source*: http://en.wikipedia.org/w/index.php?title=File:Asteroid_2004_FH.gif *License*: unknown *Contributors*: Images obtained by Stefano Sposetti, Switzerland on March 18, 2004. Animation made Raoul Behrend, Geneva Observatory, Switzerland.

Image:Moon and Asteroids 1 to 10.svg *Source*: http://en.wikipedia.org/w/index.php?title=File:Moon_and_Asteroids_1_to_10.svg *License*: unknown *Contributors*: User:Vystrix Nexoth

Image:Ceres optimized.jpg *Source*: http://en.wikipedia.org/w/index.php?title=File:Ceres_optimized.jpg *License*: unknown *Contributors*: NASA, ESA, J. Parker (Southwest Research Institute), P. Thomas (Cornell University), and L. McFadden (University of Maryland, College Park)

Image:masses of asteroids vs main belt.png *Source*: http://en.wikipedia.org/w/index.php?title=File:Masses_of_asteroids_vs_main_belt.png *License*: unknown *Contributors*: User:Kwamikagami

Image:(253) mathilde.jpg *Source*: http://en.wikipedia.org/w/index.php?title=File:(253)_mathilde.jpg *License*: unknown *Contributors*: NASA

Image:433eros.jpg *Source*: http://en.wikipedia.org/w/index.php?title=File:433eros.jpg *License*: unknown *Contributors*: NASA/NEAR Project (JHU/APL).

File:Dawn-image-070911.jpg *Source*: http://en.wikipedia.org/w/index.php?title=File:Dawn-image-070911.jpg *License*: unknown *Contributors*: NASA/JPL-Caltech/UCLA/MPS/DLR/IDA

Image:TheTransneptunians 73AU.svg *Source*: http://en.wikipedia.org/w/index.php?title=File:TheTransneptunians_73AU.svg *License*: unknown *Contributors*: User:Eurocommuter

File:Kepler orbits.svg *Source*: http://en.wikipedia.org/w/index.php?title=File:Kepler_orbits.svg *License*: unknown *Contributors*: User:Stamcose

File:Eccentricity rocky planets.jpg *Source*: http://en.wikipedia.org/w/index.php?title=File:Eccentricity_rocky_planets.jpg *License*: unknown *Contributors*: frankuitaalst from the Gravity Simulator message board.

File:Orbit1.svg *Source*: http://en.wikipedia.org/w/index.php?title=File:Orbit1.svg *License*: unknown *Contributors*: Lasunncty (talk).

Image:Orbit1.svg *Source*: http://en.wikipedia.org/w/index.php?title=File:Orbit1.svg *License*: unknown *Contributors*: Lasunncty (talk).

GNU Free Documentation License Version 1.2, November 2002 Copyright (C) 2000,2001,2002 Free Software Foundation, Inc. 59 Temple Place, Suite 330, Boston, MA 02111-1307 USA Everyone is permitted to copy and distribute verbatim copies of this license document, but changing it is not allowed.

0. PREAMBLE

The purpose of this License is to make a manual, textbook, or other functional and useful document "free" in the sense of freedom: to assure everyone the effective freedom to copy and redistribute it, with or without modifying it, either commercially or noncommercially. Secondarily, this License preserves for the author and publisher a way to get credit for their work, while not being considered responsible for modifications made by others. This License is a kind of "copyleft", which means that derivative works of the document must themselves be free in the same sense. It complements the GNU General Public License, which is a copyleft license designed for free software. We have designed this License in order to use it for manuals for free software, because free software needs free documentation: a free program should come with manuals providing the same freedoms that the software does. But this License is not limited to software manuals; it can be used for any textual work, regardless of subject matter or whether it is published as a printed book. We recommend this License principally for works whose purpose is instruction or reference.

1. APPLICABILITY AND DEFINITIONS

This License applies to any manual or other work, in any medium, that contains a notice placed by the copyright holder saying it can be distributed under the terms of this License. Such a notice grants a world-wide, royalty-free license, unlimited in duration, to use that work under the conditions stated herein. The "Document", below, refers to any such manual or work. Any member of the public is a licensee, and is addressed as "you". You accept the license if you copy, modify or distribute the work in a way requiring permission under copyright law. A "Modified Version" of the Document means any work containing the Document or a portion of it, either copied verbatim, or with modifications and/or translated into another language. A "Secondary Section" is a named appendix or a front-matter section of the Document that deals exclusively with the relationship of the publishers or authors of the Document to the Document's overall subject (or to related matters) and contains nothing that could fall directly within that overall subject. (Thus, if the Document is in part a textbook of mathematics, a Secondary Section may not explain any mathematics.) The relationship could be a matter of historical connection with the subject or with related matters, or of legal, commercial, philosophical, ethical or political position regarding them. The "Invariant Sections" are certain Secondary Sections whose titles are designated, as being those of Invariant Sections, in the notice that says that the Document is released under this License. If a section does not fit the above definition of Secondary then it is not allowed to be designated as Invariant. The Document may contain zero Invariant Sections. If the Document does not identify any Invariant Sections then there are none. The "Cover Texts" are certain short passages of text that are listed, as Front-Cover Texts or Back-Cover Texts, in the notice that says that the Document is released under this License. A Front-Cover Text may be at most 5 words, and a Back-Cover Text may be at most 25 words. A "Transparent" copy of the Document means a machine-readable copy, represented in a format whose specification is available to the general public, that is suitable for revising the document straightforwardly with generic text editors or (for images composed of pixels) generic paint programs or (for drawings) some widely available drawing editor, and that is suitable for input to text formatters or for automatic translation to a variety of formats suitable for input to text formatters. A copy made in an otherwise Transparent file format whose markup, or absence of markup, has been arranged to thwart or discourage subsequent modification by readers is not Transparent. An image format is not Transparent if used for any substantial amount of text. A copy that is not "Transparent" is called "Opaque". Examples of suitable formats for Transparent copies include plain ASCII without markup, Texinfo input format, LaTeX input format, SGML or XML using a publicly available DTD, and standard-conforming simple HTML, PostScript or PDF designed for human modification. Examples of transparent image formats include PNG, XCF and JPG. Opaque formats include proprietary formats that can be read and edited only by proprietary word processors, SGML or XML for which the DTD and/or processing tools are not generally available, and the machine-generated HTML, PostScript or PDF produced by some word processors for output purposes only. The "Title Page" means, for a printed book, the title page itself, plus such following pages as are needed to hold, legibly, the material this License requires to appear in the title page. For works in formats which do not have any title page as such, "Title Page" means the text near the most prominent appearance of the work's title, preceding the beginning of the body of the text. A section "Entitled XYZ" means a named subunit of the Document whose title either is precisely XYZ or contains XYZ in parentheses following text that translates XYZ in another language. (Here XYZ stands for a specific section name mentioned below, such as "Acknowledgements", "Dedications", "Endorsements", or "History".) To "Preserve the Title" of such a section when you modify the Document means that it remains a section "Entitled XYZ" according to this definition. The Document may include Warranty Disclaimers next to the notice which states that this License applies to the Document. These Warranty Disclaimers are considered to be included by reference in this License, but only as regards disclaiming warranties: any other implication that these Warranty Disclaimers may have is void and has no effect on the meaning of this License.

2. VERBATIM COPYING

You may copy and distribute the Document in any medium, either commercially or noncommercially, provided that this License, the copyright notices, and the license notice saying this License applies to the Document are reproduced in all copies, and that you add no other conditions whatsoever to those of this License. You may not use technical measures to obstruct or control the reading or further copying of the copies you make or distribute. However, you may accept compensation in exchange for copies. If you distribute a large enough number of copies you must also follow the conditions in section 3. You may also lend copies, under the same conditions stated above, and you may publicly display copies.

3. COPYING IN QUANTITY

If you publish printed copies (or copies in media that commonly have printed covers) of the Document, numbering more than 100, and the Document's license notice requires Cover Texts, you must enclose the copies in covers that carry, clearly and legibly, all these Cover Texts: Front-Cover Texts on the front cover, and Back-Cover Texts on the back cover. Both covers must also clearly and legibly identify you as the publisher of these copies. The front cover must present the full title with all words of the title equally prominent and visible. You may add other material on the covers in addition. Copying with changes limited to the covers, as long as they preserve the title of the Document and satisfy these conditions, can be treated as verbatim copying in other respects. If the required texts for either cover are too voluminous to fit legibly, you should put the first ones listed (as many as fit reasonably) on the actual cover, and continue the rest onto adjacent pages. If you publish or distribute Opaque copies of the Document numbering more than 100, you must either include a machine-readable Transparent copy along with each Opaque copy, or state in or with each Opaque copy a computer-network location from which the general network-using public has access to download using public-standard network protocols a complete Transparent copy of the Document, free of added material. If you use the latter option, you must take reasonably prudent steps, when you begin distribution of Opaque copies in quantity, to ensure that this Transparent copy will remain thus accessible at the stated location until at least one year after the last time you distribute an Opaque copy (directly or through your agents or retailers) of that edition to the public. It is requested, but not required, that you contact the authors of the Document well before redistributing any large number of copies, to give them a chance to provide you with an updated version of the Document.

4. MODIFICATIONS

You may copy and distribute a Modified Version of the Document under the conditions of sections 2 and 3 above, provided that you release the Modified Version under precisely this License, with the Modified Version filling the role of the Document, thus licensing distribution and modification of the Modified Version to whoever possesses a copy of it. In addition, you must do these things in the Modified Version: A. Use in the Title Page (and on the covers, if any) a title distinct from that of the Document, and from those of previous versions (which should, if there were any, be listed in the History section of the Document). You may use the same title as a previous version if the original publisher of that version gives permission. B. List on the Title Page, as authors, one or more persons or entities responsible for authorship of the modifications in the Modified Version, together with at least five of the principal authors of the Document (all of its principal authors, if it has fewer than five), unless they release you from this requirement. C. State on the Title page the name of the publisher of the Modified Version, as the publisher. D. Preserve all the copyright notices of the Document. E. Add an appropriate copyright notice for your modifications adjacent to the other copyright notices. F. Include, immediately after the copyright notices, a license notice giving the public permission to use the Modified Version under the terms of this License, in the form shown in the Addendum below. G. Preserve in that license notice the full lists of Invariant Sections and required Cover Texts given in the Document's license notice. H. Include an unaltered copy of this License. I. Preserve the section Entitled "History", Preserve its Title, and add to it an item stating at least the title, year, new authors, and publisher of the Modified Version as given on the Title Page. If there is no section Entitled "History" in the Document, create one stating the title, year, authors, and publisher of the Document as given on its Title Page, then add an item describing the Modified Version as stated in the previous sentence. J. Preserve the network location, if any, given in the Document for public access to a Transparent copy of the Document, and likewise the network locations given in the Document for previous versions it was based on. These may be placed in the "History" section. You may omit a network location for a work that was published at least four years before the Document itself, or if the original publisher of the version it refers to gives permission. K. For any section Entitled "Acknowledgements" or "Dedications", Preserve the Title of the section, and preserve in the section all the substance and tone of each of the contributor acknowledgements and/or dedications given therein. L. Preserve all the Invariant Sections of the Document, unaltered in their text and in their titles. Section numbers or the equivalent are not considered part of the section titles. M. Delete any section Entitled "Endorsements". Such a section may not be included in the Modified Version. N. Do not retitle any existing section to be Entitled "Endorsements" or to conflict in title with any Invariant Section. O. Preserve any Warranty Disclaimers. If the Modified Version includes new front-matter sections or appendices that qualify as Secondary Sections and contain no material copied from the Document, you may at your option designate some or all of these sections as invariant. To do this, add their titles to the list of Invariant Sections in the Modified Version's license notice. These titles must be distinct from any other section titles. You may add a section Entitled "Endorsements", provided it contains nothing but endorsements of your Modified Version by various parties--for example, statements of peer review or that the text has been approved by an organization as the authoritative definition of a standard. You may add a passage of up to five words as a Front-Cover Text, and a passage of up to 25 words as a Back-Cover Text, to the end of the list of Cover Texts in the Modified Version. Only one passage of Front-Cover Text and one of Back-Cover Text may be added by (or through arrangements made by) any one entity. If the Document already includes a cover text for the same cover, previously added by you or by arrangement made by the same entity you are acting on behalf of, you may not add another; but you may replace the old one, on explicit permission from the previous publisher that added the old one. The author(s) and publisher(s) of the Document do not by this License give permission to use their names for publicity for or to assert or imply endorsement of any Modified Version.

5. COMBINING DOCUMENTS

You may combine the Document with other documents released under this License, under the terms defined in section 4 above for modified versions, provided that you include in the combination all of the Invariant Sections of all of the original documents, unmodified, and list them all as Invariant Sections of your combined work in its license notice, and that you preserve all their Warranty Disclaimers. The combined work need only contain one copy of this License, and multiple identical Invariant Sections may be replaced with a single copy. If there are multiple Invariant Sections with the same name but different contents, make the title of each such section unique by adding at the end of it, in parentheses, the name of the original author or publisher of that section if known, or else a unique number. Make the same adjustment to the section titles in the list of Invariant Sections in the license notice of the combined work. In the combination, you must combine any sections Entitled "History" in the various original documents, forming one section Entitled "History"; likewise combine any sections Entitled "Acknowledgements", and any sections Entitled "Dedications". You must delete all sections Entitled "Endorsements".

6. COLLECTIONS OF DOCUMENTS

You may make a collection consisting of the Document and other documents released under this License, and replace the individual copies of this License in the various documents with a single copy that is included in the collection, provided that you follow the rules of this License for verbatim copying of each of the documents in all other respects. You may extract a single document from such a collection, and distribute it individually under this License, provided you insert a copy of this License into the extracted document, and follow this License in all other respects regarding verbatim copying of that document.

7. AGGREGATION WITH INDEPENDENT WORKS

A compilation of the Document or its derivatives with other separate and independent documents or works, in or on a volume of a storage or distribution medium, is called an "aggregate" if the copyright resulting from the compilation is not used to limit the legal rights of the compilation's users beyond what the individual works permit. When the Document is included in an aggregate, this License does not apply to the other works in the aggregate which are not themselves derivative works of the Document. If the Cover Text requirement of section 3 is applicable to these copies of the Document, then if the Document is less than one half of the entire aggregate, the Document's Cover Texts may be placed on covers that bracket the Document within the aggregate, or the electronic equivalent of covers if the Document is in electronic form. Otherwise they must appear on printed covers that bracket the whole aggregate.

8. TRANSLATION

Translation is considered a kind of modification, so you may distribute translations of the Document under the terms of section 4. Replacing Invariant Sections with translations requires special permission from their copyright holders, but you may include translations of some or all Invariant Sections in addition to the original versions of these Invariant Sections. You may include a translation of this License, and all the license notices in the Document, and any Warranty Disclaimers, provided that you also include the original English version of this License and the original versions of those notices and disclaimers. In case of a disagreement between the translation and the original version of this License or a notice or disclaimer, the original version will prevail. If a section in the Document is Entitled "Acknowledgements", "Dedications", or "History", the requirement (section 4) to Preserve its Title (section 1) will typically require changing the actual title.

9. TERMINATION

You may not copy, modify, sublicense, or distribute the Document except as expressly provided for under this License. Any other attempt to copy, modify, sublicense or distribute the Document is void, and will automatically terminate your rights under this License. However, parties who have received copies, or rights, from you under this License will not have their licenses terminated so long as such parties remain in full compliance.

10. FUTURE REVISIONS OF THIS LICENSE

The Free Software Foundation may publish new, revised versions of the GNU Free Documentation License from time to time. Such new versions will be similar in spirit to the present version, but may differ in detail to address new problems or concerns. See http://www.gnu.org/copyleft/. Each version of the License is given a distinguishing version number. If the Document specifies that a particular numbered version of this License "or any later version" applies to it, you have the option of following the terms and conditions either of that specified version or of any later version that has been published (not as a draft) by the Free Software Foundation. If the Document does not specify a version number of this License, you may choose any version ever published (not as a draft) by the Free Software Foundation. ADDENDUM: How to use this License for your documents To use this License in a document you have written, include a copy of the License in the document and put the following copyright and license notices just after the title page: Copyright (c) YEAR YOUR NAME. Permission is granted to copy, distribute and/or modify this document under the terms of the GNU Free Documentation License, Version 1.2 or any later version published by the Free Software Foundation; with no Invariant Sections, no Front-Cover Texts, and no Back-Cover Texts. A copy of the license is included in the section entitled "GNU Free Documentation License". If you have Invariant Sections, Front-Cover Texts and Back-Cover Texts, replace the "with...Texts." line with this: with the Invariant Sections being LIST THEIR TITLES, with the Front-Cover Texts being LIST, and with the Back-Cover Texts being LIST. If you have Invariant Sections without Cover Texts, or some other combination of the three, merge those two alternatives to suit the situation. If your document contains nontrivial examples of program code, we recommend releasing these examples in parallel under your choice of free software license, such as the GNU General Public License, to permit their use in free software.

MIX
Papier aus verantwortungsvollen Quellen
Paper from responsible sources
FSC® C105338

Printed by Books on Demand GmbH, Norderstedt / Germany